EXTINCTION

BAD GENES

OR BAD LUCK?

EXTINCTION

BAD GENES

OR BAD LUCK?

DAVID M. RAUP

W·W·NORTON & COMPANY
New York · London

The text of this book is composed in Bembo.
Composition by the Haddon Craftsmen, Inc.
Book design by Jacques Chazaud.

Library of Congress Cataloging-in-Publication Data

Raup, David M.
    Extinction : bad genes or bad luck? / David M. Raup.
      p. cm.
    Includes index.
    1. Extinction (Biology)   I. Title.
  QH78.R38 1991
  575'.7—dc20 90-27192

ISBN 0-393-03008-3

W.W. Norton & Company, Inc., 500 Fifth Avenue, New York, N.Y. 10110
W.W. Norton & Company, Ltd., 10 Coptic Street, London WC1A 1PU

CONTENTS

v

CONTENTS

ILLUSTRATIONS

ix

PREFACE

This is a book about the history of life on earth: the myriad twistings and turnings that have led, somehow, to us. It is written in the conviction that our biological origins are at least as important—and as interesting—as the physical origins of our universe.

My emphasis throughout is on extinction—the death of species—a facet of organic evolution that has received surprisingly little attention. The main question, to be visited again and again, is whether the billions of species that died in the geologic past died because they were less fit (bad genes) or merely because they were in the wrong place at the wrong time (bad luck). Do species struggle or gamble for survival? This leads to a question closer to home: Are we here because of a natural superiority (opposable thumbs, big brains, and so on), or are we just plain lucky? In other words, is the evolution of life a fair game, as the survival-of-the-fittest doctrine so strongly implies?

Apart from the inherent importance of past extinctions, the subject also bears on the contemporary problems of endangered species, losses of biodiversity, and extinctions caused by human activities. The history of species extinction provides valuable perspective on global ecology of the present and the future.

It is a pleasure to thank the people and institutions that helped make this book possible. NASA's Exobiology program has supported my research on extinction for several years, as part of the agency's continuing investigation of life in the universe, and I am grateful for this support. The University of Chicago provided the intellectual atmosphere and challenging students so necessary to the enterprise of trying to think about old problems in new ways.

I am indebted to my colleague Jack Sepkoski for making freely available his massive compilations of data on extinction. I also thank Jack and our mutual colleague Dave Jablonski for countless discussions of the extinction phenomenon. We have knocked these problems around so often that it is impossible to say which ideas came from whom. Neither Jack nor Dave, however, should be held responsible for some of the more bizarre ideas expressed in the pages that follow.

On the production side, I am grateful to my wife, Judie Yamamoto, for her continued support and tolerance and for reading the manuscript in draft after draft. Vital and much appreciated has been Wesley Gray's help in keeping the computer going. Also, my thanks to Clark Chapman, Marianne Fons, Kubet Luchterhand, Daniel McShea, Matthew Nitecki, Jack Sepkoski, and Gene Shoemaker, who read the manuscript and offered many good suggestions. Finally, Ed Barber at Norton has supported the project throughout; his editing was constructive and much appreciated.

Chicago
November 1990

INTRODUCTION

Stephen Jay Gould
MUSEUM OF COMPARATIVE ZOOLOGY
HARVARD UNIVERSITY

Nature and the earth are traditionally personified as female; but all professions and institutions (as a commentary both on our personal need for continuity and on the sexism of our society) have conventional fathers—Washington of his country, Copernicus of modern astronomy. Charles Lyell is the acknowledged "father" of geology, primarily for a codification of modernity in his multivolume textbook, *Principles of Geology* (1830–33). Lyell encapsulated his philosophy in a doctrine later called "uniformitarianism"—a complex set of beliefs centered on the catechism that "the present is the key to the past." Lyell argued that the current range of natural processes, and their rates of action, would be sufficient, in principle, to account for the entire panoply of past causes in the full history of the earth and life. Lyell

xiii

viewed this principle as a methodological reform to eliminate fanciful (and quasi-theological) "catastrophic" causes and to render the full magnitude of past change by the slow and steady accumulation of ordinary small changes (deposition and erosion grain by grain) extended over vast times.

The idea sounds so sensible and right-minded. Even the greatest modern revolution in geology, continental drift and plate tectonics, embodies the uniformitarian view in its vision of the efficacy of continental motion at a few centimeters per year extrapolated to great cumulative change over the immensity of geologic time. And yet, from two different standpoints (theoretical and empirical), Lyell's credo makes little sense, and its status as dogma can only reflect our social and psychological preferences. First, what is the probability that our tiny slice of observable time should include the full range of potential processes that might alter the earth? What about big, but perfectly natural, events that occur so infrequently that we have only a remote chance of observing even one occurrence in historical time? Second, how can Lyellian gradualism account for the fundamental fact of paleontology—extensive, and apparently rapid, faunal turnovers ("mass extinctions") occurring several times in the history of life? (Traditional explanations try to "smooth out" these big dyings by extending them over at least a few million years and attributing them to intensification of ordinary causes—changes in temperature and sea level, for example—but the arguments have always seemed forced.)

The extraterrestrial-impact theory of mass extinction— easy to conceptualize, but proposed with good supporting data for the first time only in 1980—may revolutionize not only our view of life's history but our entire concept of historical change by making a legitimate catastrophism respectable again. Thus, as the historian of science William Glen has argued, theories about the efficacy of impact may be conceptually more important and farther ranging than even plate tectonics, which reformed our view of planetary mechanics but left the Lyellian view of change unscathed.

The impact scenario is now well documented only for the Cretaceous extinction—not the largest by any means, but particularly dear to our hearts because it wiped out dinosaurs and gave mammals a chance, thereby making our own evolution possible. But an extension to other mass extinctions becomes an exciting possibility and a hot subject for current research. The fossil record is our wellspring of data for this subject, and the quantification of patterns is extinction is the most important and promising subject of all. Yet, until recently, extinction received much less attention than its obvious prominence warranted. In an overly Darwinian world of adaptation, gradual change, and improvement, extinction seemed, well, so negative—the ultimate failure, the flip side of evolution's "real" work, something to be acknowledged but not intensely discussed in polite company.

This odd neglect has been reversed in the last decade, and no topic now commands more interest among paleontologists than extinction. The reasons are many, with a prominent root to the impact theory of mass dying. But the primary architect of this shift is my brilliant colleague David M. Raup. Dave may be more at home before a computer console than before a dusty drawer of fossils (and he gets his share of flak from traditionalists for this predilection), but he is the acknowledged master of quantitative approaches to the fossil record. He saw the power of the impact scenario right from the beginning, when most paleontologists were howling with rage or laughter, and refusing to consider the proposal seriously. He has made the most important discoveries and proposed the most interesting and outrageous hypotheses in the field, including the suggestion that mass extinctions may cycle with a frequency of some 26 million years. He is also the perennial Peck's bad boy of paleontology—a hard act to maintain past the age of fifty (I am struggling with him), but truly the most sublime of all statuses in science. If Dave has any motto, it can only be: Think the unthinkable (and then make a mathematical model to show how it might work); take an outrageous idea with a limited sphere of validity and see if it might not be extendable to explain everything. This book is a wonderful exposition of this potentially valid iconoclasm, as Dave not only validates the impact scenario for one major extinction but then asks, Suppose that all extinctions, not just mass dyings but even minor removals in local areas, are caused by impacts of varying sizes. What would the history of life look like then? Does the actual history of life look like this after all?

Paleontology, though imbued with the usual swirling debate so characteristic of all interesting science, is a relatively friendly profession. I like almost all my colleagues, but I reserve special affection for a handful who have inspired me by their insight and pushed me to consider outrageous novelty. Dave Raup is the best of the best. Before we met personally, he reviewed the first paper I ever wrote and, with a maximum of friendly encouragement to a (then) very insecure graduate student, showed me significant errors in the most supportive way. We worked together on a long series of papers in the 1970s and 1980s (on the power of

random processes to produce an overt order that most paleontologists would misinterpret as prima facie evidence of
conventional causation). Dave was then toying with the
idea that all mass extinctions are artifacts, that underlying
rates never vary through time, and that apparent wipeouts
are false appearances based upon an imperfect fossil record.
("But Dave," I used to remonstrate, "you just can't mean it.
The issue isn't that 95 percent of species die at the end of the
Permian. The main point is that they never come back again
in the Triassic—so they must have really died." I think I
won that argument, but the colloquy does prove that Dave
will consider anything and can change his mind.)

It is fashionable these days to consider jealousy, conniv-
ing, and fraud as norms of interaction among scientists. But
this dubious behavior seems to prevail only because malfeas-
ance is so much more visible than simple kindness, helpful-
ness, and collegiality. Ten thousand acts of kindness, never
recorded, more than balance every headline for single items
of venality. This collegiality is the glue of science and the
source of delight in a profession that otherwise abounds in
tedium and the insane pressure of multifarious commit-
ments. I rejoice that I have colleagues like Dave, people
with unshakable integrity and probing wit. With friends
and prods like this, we will never grow old or cynical.

EXTINCTION

BAD GENES

OR BAD LUCK?

CHAPTER **I**

ALMOST ALL SPECIES
ARE EXTINCT

Almost all professional football players are still alive. The same is probably true for nuclear physicists, city planners, and tax consultants. This survival record is due in part to the newness of professional football, nuclear physics, and so on and in part to population growth—there are lots more people now than ever before. Nathan Keyfitz, the great demographer, estimated in 1966 that about 4 percent of all the people who had ever lived were alive then. Again, newness and population growth.

Not so for species! There are millions of different species of animals and plants on earth—possibly as many as forty million. But somewhere between five and fifty *billion* species have existed at one time or another. Thus, only about one in a thousand species is still alive—a truly lousy survival record: 99.9 percent failure! This book examines two primary questions: Why did so many species die out? How did they die out?

Is Extinction Important?

Yes, I think it is very important. All of us grow up with an acquired set of ideas and thoughts about the natural world around us, its history and its future. We get these ideas from a thousand sources—from comic books and classrooms and television sitcoms—and these ideas represent the collective attitudes of our culture. One idea I think most of us share is that earth is a pretty safe and benevolent place to live—not counting what humans can do to earth and to each other. Earthquakes, hurricanes, and disease epidemics may strike, but on the whole our planet is stable. It is neither too warm nor too cold, the seasons are predictable, and the sun rises and sets on schedule.

Much of our good feeling about planet earth stems from a certainty that life has existed without interruption for three and a half billion years. We have been taught, as well, that most changes in the natural world are slow and gradual. Species evolve in tiny steps over eons; erosion and weathering change our landscape but at an almost immeasurably slow pace. Continents move, as in the present drift of North America away from Europe, but this movement is measured in centimeters per year and will have no practical effect on our lives or those of our children.

Is all this true or merely a fairy tale to comfort us? Is there more to it? I think there is. Almost all species in the past failed. If they died out gradually and quietly and if they

4

deserved to die because of some inferiority, then our good feelings about earth can remain intact. But if they died violently and without having done anything wrong, then our planet may not be such a safe place.

BAD GENES OR BAD LUCK?

I have taken the title of this book from a research article I published in Spain some years ago. I was concerned then with the failure of trilobites in the Paleozoic era. Starting about 570 million years ago, these complex, crab-like organisms dominated life on ocean bottoms—at least they dominate the fossil assemblages of that age. But through the 325 million years of the Paleozoic era, trilobites dwindled in numbers and variety, finally disappearing completely in the mass extinction that ended the era, about 245 million years ago. As far as we know, trilobites left no descendants.

My question in Spain is the one I still ask: Why? Did the trilobites do something wrong? Were they fundamentally inferior organisms? Were they stupid? Or did they just have the bad luck to be in the wrong place at the wrong time? The first alternative, bad genes, could be manifested by things like susceptibility to disease, lack of good sensory perception, or poor reproductive capacity. The second, bad luck, could be a freak catastrophe that eliminated all life in areas where trilobites happened to be living. The question is basically one of nature versus nurture. Is proneness to extinction an inherent property of a species—a weakness—or does it depend on vagaries of chance in a risk-ridden world?

Of course, the problem is more complex than I have presented it, just as the nature-nurture question in human behavior is complex. But in both situations, nature (genetics) and nurture (environment) operate to some degree, and the challenge is to find out which process dominates and whether the imbalance varies in time and space.

THE NATURE OF EXTINCTION

We could avoid the extinction problem—and therefore this book—by arguing it away. We could note that the average plant or animal species has a geologic life span of only four million years and that life goes back thousands of millions of years. On this basis, we could assert that it is nature's way for species to have short life spans. Will Cuppy, in his delightful volume of essays entitled *How to Become Extinct,* wrote, "The Age of Reptiles ended because it had gone on long enough and it was all a mistake in the first place."

If we accept that turnover in species is merely nature's way, just as nature has given humans a limited life span, then there is nothing in species extinction worthy of wonder. But there is absolutely no basis for equating the life spans of species with those of individual humans. There is no evidence of aging in species or any known reason why a species could not live forever. In fact, virtual immortality has been claimed for the so-called living fossils (cockroaches and sharks, for example).

We could also get rid of the problem by arguing that species don't go extinct, that they merely evolve into other species (presumably better ones) by natural selection. The essence of Darwin's *Origin of Species* was that species change gradually into new species. When a new species is formed in

6

this way, the ancestral species does not die: it is merely transformed into another species. The ancestral species is then said to have undergone "pseudoextinction"—as opposed to "true extinction." Although pseudoextinction certainly occurs in nature, we also know that true extinction has eliminated countless species. Many major groups of plants and animals that were once important parts of the global biota no longer exist and left no descendants. Much of the debate in evolutionary biology over the "punctuated equilibrium" theory (championed by Stephen Jay Gould) has centered on the question of proportions of true and pseudo extinctions in the history of life.

Another kind of false extinction has been claimed. It has been argued that dinosaurs did not die out, but just evolved wings and flew away. At a certain level, this reasoning is sound. Birds evolved in the Jurassic period, about 150 million years ago, from dinosaurs of the time (Figure 1-1). The first fossil birds are hardly distinguishable from some smaller Jurassic dinosaurs. So birds, as a group, did descend from dinosaurs and have many anatomical similarities to show for it. All 8,600 species of birds living today carry some inheritance from their reptilian ancestors.

But the bird lineage split off millions of years before the dinosaurs died out in the mass extinction that ended the Cretaceous period. Cretaceous dinosaurs died without issue! Their extinction was final. We cannot escape the fact that true extinction has claimed a large fraction of the evolutionary progeny of life on earth—even though the size of that fraction is not known accurately.

Evolution of Birds from Jurassic Dinosaurs

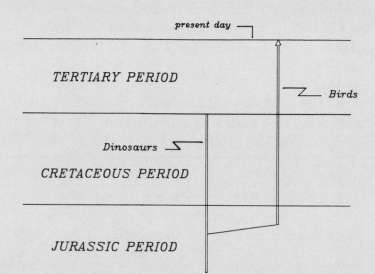

FIGURE 1-1.   Evolutionary tree (greatly simplified) showing the origin of birds from the dinosaur lineage in the Jurassic period. This pattern has led some to argue that dinosaurs did not die out at the end of the Cretaceous, that they merely evolved wings in the Jurassic and flew away.

WHO STUDIES EXTINCTION?

Strangely, extinction does not have a large body of scholars or scholarship. No scientific discipline carries the name. Nevertheless, we know a lot about the subject. Early in the nineteenth century, geologists discovered that short durations of fossil species provide the best means of arranging

8

geologic events in a time sequence. Reversing the reasoning, geologists can locate a rock in time with great precision by identifying the fossils. Even today, much of the world's exploration for oil and natural gas is built around a chronology based on a changing fossil record—in other words, origination and extinction of species.

But the geologists, and their colleagues in paleontology who have done most of the hard work, have never developed a strong interest in extinction itself. It may be that these scientists operate so close to the fossil record that they have lost any natural feelings of awe. Since virtually all species found in rocks are extinct, the question becomes not why but merely when. Surprisingly few geologists and paleontologists have taken an active part in today's concern about endangered species and predictions of future extinctions.

When I was in graduate school training to be a paleontologist, I did learn a few things about extinction. I learned that species compete constantly with each other for space and resources and struggle always with their physical environment. I learned that a steady, background level of extinction is an inevitable part of the history of life, punctuated occasionally by big events called mass extinctions. Conventional wisdom went little further. Although some attention in classes and textbooks was devoted to mass extinctions, these events were seen as too complex ever to be understood. Our mission in school was to learn to identify the most important fossils and their ranges in geologic time.

But if geologists and paleontologists don't pay serious attention to extinction, surely the biologists care. Organic evolution is central to almost all aspects of biology. Whole subdisciplines—molecular and population genetics, taxonomy, and many areas of ecology and biogeography—seek to document evolutionary history or to investigate the process by which organisms evolve. Who begat whom, and when, why, and how? But to the typical biologist, extinction plays a strangely minor role in evolution.

An important topic in biology of the past several decades has been a phenomenon known as *speciation*. By common agreement, the term refers to the splitting or branching of an evolutionary line, producing two distinct species where there was one. Ironically, this is not the origin of species that Charles Darwin focused on. The main mode of Darwinian change is the gradual transformation of one species into another—with no increase in the number of coexisting species. In fact, the Darwinian kind of species origin is not even called speciation by most biologists, but rather, and somewhat awkwardly, *phyletic transformation*.

In Figure 1-2, I have illustrated the difference between speciation and phyletic transformation. The two lines with symbols attached depict species lineages of some imaginary beast evolving through time. The anatomy of one species is caricatured by circles. With time, the circles become smaller (on average), suggesting evolution toward smaller body size. This change occurs by phyletic transformation.

Partway through the sequence in Figure 1-2, a branching occurs: some of the circular organisms split off to start a lineage of square organisms. The squares then evolve by phyletic transformation; in this case, the anatomical change happens to be toward size increase. Note that I have made the ancestral species (round) go extinct but have let the descendant species (square) live on.

If 99.9 percent of all species that have lived on earth are

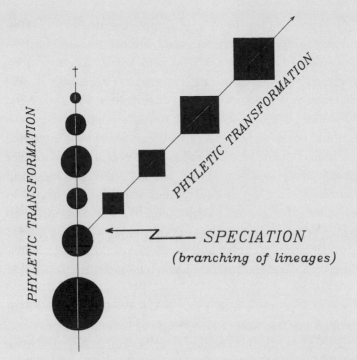

FIGURE 1-2. Hypothetical evolutionary tree showing the distinction between phyletic transformation and speciation. Imaginary organisms (circles and squares) change gradually through time, becoming larger or smaller (PHYLETIC TRANSFORMATION). At the branch point (SPECIATION), the round organisms give rise to a lineage of square organisms.

extinct, it follows that the total of species originations has been virtually the same as the total of species extinctions. Although present biodiversity—the millions of living species—seems high to us, today's biota results from a minor surplus of speciations over extinctions, accumulated over a long time.

In view of these figures, it is puzzling that even evolutionary biologists have devoted almost no attention to extinction. Large monographs and textbooks have been written about speciation, and careers have been built around the subject. But extinction has barely been touched. It's a little like a demographer trying to study population growth without considering death rates. Or an accountant interested in credits but not debits. Textbooks of evolutionary biology contain little about extinction beyond a few platitudes and tautologies like "Species go extinct when they are unable to cope with change" or "Extinction is likely when population size approaches zero." The *Encyclopaedia Britannica* (1987) says, "Extinction occurs when a species can no longer reproduce at replacement levels." These statements are almost free of content.

But interests in science change, and this is happening with extinction. Thanks to a rash proposal by a Nobel physicist, Luis Alvarez, and his colleagues at Berkeley, furious debate has broken out over whether a meteorite impact caused the dinosaur extinction. This has combined with concern over currently endangered species to encourage more people to probe the extinction phenomenon and its role in the history of life. The emerging field of extinction studies may one day even get a name ending in "ology." My mission in this book is to share with you the attempts being made by many of us to understand more about extinction.

I should emphasize that extinction is still a very small, cottage industry. It has none of the trappings of big science—nothing comparable to the Supercollider or the Human Genome Project or the Hubble Space Telescope. Yet the questions being asked of extinction are every bit as fundamental and interesting in our ongoing attempt to understand our place in the universe and to answer the ultimate question: Why are we here?

A Word about the Word

Curiously, the word *extinct* is an adjective. We say that species (or volcanoes) *become* or *go* extinct. The word has a passive quality, implying a petering out. Although *extinct* once was used as an active verb, this usage passed out of the English language in the seventeenth century. Plants and animals do all kinds of active things: they fight, eat, migrate, reproduce, and even speciate. But when species die, they *become* extinct. Perhaps extinction, as the death of species, is a little scary and we unconsciously avoid the active voice. Or perhaps the usage is meant to imply that the species becoming extinct is reacting to outside influences, beyond its control. This would be reasonable, I suppose, because there is no reason to suspect that any species is actively suicidal, even though some of its members may be.

Digby McLaren, a noted Canadian paleontologist and student of extinction, argues for the use of the term *mass killing* in place of *mass extinction.* But he has done this to distinguish death of individual animals from death of species. McLaren is convinced that the most dramatic aspect of mass extinctions is sudden killing of a multitude of individuals. Extinction of a species, to him, is merely a byproduct in cases where the killing happens to be complete. Thus, McLaren is not suggesting a change in language—only a change of emphasis from species to individuals.

In a couple of recent research articles, I have gone yet further by actually using *kill* and *killing* in place of *extinct* and *extinction*. I am waiting, somewhat impishly, to see whether this usage is picked up by my colleagues. At the very least, I expect it will lead to some delightful conversations with Digby McLaren.

SPECIES DEFINED

Before going much further, I should clarify what I mean by *species*. The species is the traditional unit of accounting in most extinction studies, McLaren's views notwithstanding.

A species is a species if a competent taxonomist says it is. Although a bit cynical, this is the operational definition most widely used in biology and paleontology. It works because the biological world is, in fact, divided into natural units. Professional taxonomists devote much time and energy to classifying the organic world into its basic units— kinds of organisms distinguishable from other kinds. Criteria include anatomy, biochemistry, color, breeding systems, and sometimes behavior. The taxonomist's experience is used to choose characteristics that make consistent classification possible.

A more rigorous definition is possible: *a species is a group of individual organisms that share a common pool of genetic material (genome).* All humans belong to a single species because they are all interfertile. Aside from gender itself, the only barriers to reproduction among members of our species are geographic and cultural. The biological world is an array of separate and independent genomes, each changing through time but not mixing with others. Because species are reproductively isolated, differences in anatomy and behavior evolve.

The job of the taxonomist is to recognize and distinguish natural species. Unfortunately, breeding experiments to test for reproductive isolation are usually impractical; such tests may even be impossible if the organisms live in different areas and do not behave naturally in captivity. Thus, the taxonomist usually relies on proxy information—physical appearance, behavior, breeding cycles, and so on.

The taxonomist's task is made difficult by the fact that there are differences within as well as between species. Populations of a species living in one area may be different—often strikingly so—from populations of the same species in another area. The differences may stem from minor adaptations to local conditions or simply from chance differences that develop between populations that normally do not interbreed. Geographic variants of a species are called *subspecies, varieties,* or *races,* indicating that they could interbreed if they lived in the same area (and felt like it). Subspecies are incipient species; that is, the original species is in the process of speciation. Subspecies become fully independent species if the geographic separation is maintained long enough.

There is occasional successful hybridization between species, especially in plants (oaks, for example), which tends to blur species boundaries. Hybrids are often intermediate in form. If hybridization were rampant in our world, the whole classification of organisms into species would break down. Fortunately for taxonomists, and probably for evolution as well, this has not happened. In the latter regard, it is the existence of independently evolving genomes that enables adaptations as different as flying and swimming to evolve and persist. Without the barriers, our world would be very different and we probably would not be here. Global biology would quite likely be dominated by generalized organisms that could do a little of everything—but nothing very well.

Because it is rarely possible or practical actually to test the breeding capability of organisms, the taxonomist must make many educated guesses about species boundaries. One can verify that this approach works well most of the time by comparing classifications of different taxonomists. Especially striking have been comparisons of bird species lists for remote parts of New Guinea made independently by Western ornithologists and by local native tribes. The match is almost perfect.

Paleontologists classify fossils much as biologists classify living organisms. Breeding experiments are never possible with fossils, of course, nor is there much behavioral or physiological information. But being limited to the external appearance is not a striking disadvantage when you consider that the biologist working with living organisms also makes most decisions on this basis.

THE PURPOSE OF EXTINCTION, IF ANY

Has extinction been a good thing or merely a destructive nuisance barely overridden by the constructive forces of evolution? This is an interesting and tough question without a firm answer. One common opinion is "Of course extinction is a good thing, because it serves to weed out the less fit species." This deeply embedded notion can be found throughout Darwin's *Origin of Species* even though his main emphasis was always on fitness *within* species. To some people, the idea that extinction is ultimately good is so self-evident that it does not need testing: more fit species can be distinguished from the less fit by the mere fact of their survival.

The disturbing reality is that for none of the thousands of well-documented extinctions in the geologic past do we have a solid explanation of why the extinction occurred. We have many proposals in specific cases, of course: trilobites died out because of competition from newly evolved fish; dinosaurs were too big or too stupid; the antlers of Irish elk became too cumbersome. These are all plausible scenarios, but no matter how plausible, they cannot be shown to be true beyond reasonable doubt. Equally plausible alternative scenarios can be invented with ease, and none has predictive power in the sense that it can show a priori that a given species or anatomical type was destined to go extinct.

Sadly, the only evidence we have for the inferiority of victims of extinction is the fact of their extinction—a circular argument. The weakness of the argument does not, of course, invalidate the notion that extinction is based on fitness: it may only reflect our ignorance. For example, mammals of the late Cretaceous may have actually been better adapted than dinosaurs, but our knowledge of these animals may not be good enough for us to recognize that superiority.

Consider, as a thought experiment, what evolution would be like if there were never any extinction of species.

Figure 1-3 shows sketches of two hypothetical evolutionary trees. In both, time moves from bottom to top, and each line is a species lineage. The present is the horizontal line across the top, so that species living today are indicated by those lineages that reach the line. Both trees branch upward—like

Evolutionary Trees

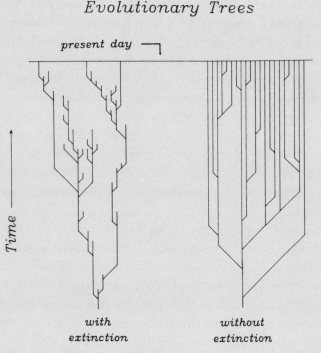

FIGURE 1-3.   Hypothetical evolutionary trees showing the effect of species extinction on biodiversity. The tree on the left reflects the actual history of life, with many species formed by lineage branching but most going extinct. Only three species survive to the present day. The tree on the right is what evolution would look like if species never went extinct: the number of coexisting species (biodiversity) would increase until saturation was reached.

bushes without a central trunk. Each branch point is an event of speciation.

The evolutionary tree on the left is what evolution actually looks like. Reading from the bottom, species lineages that terminate before reaching the top have gone extinct. The number of coexisting species (biodiversity) changes over time as new species are added by speciation and others are lost by extinction.

The tree on the right follows the same rules except that species never die. The general aspect is more like a willow than a blueberry bush. Now, we know that the tree on the left is a better description of the actual history of life, because we have ample evidence for extinction. But would evolution without extinction work?

It would probably work, but not very well. Evolution without extinction suggests several problems. Most important, biodiversity would increase exponentially. The more species lineages that came into being, the more lineages there would be to produce more species. Rather soon, the system would saturate: speciation would have to stop because there would be no room for new species.

Adaptation by natural selection would continue to hone and refine the existing species, and the ultimate quality of adaptation might even be greater than what we see today because the species would have more time. The first-formed organisms might have evolved far better structures than the ones visible today.

Thus, one can imagine an evolutionary system organized without extinction—and this may exist on planets elsewhere in space. But would an extinction-free world have produced as much biological variety as has evolved on earth—organisms as different as trilobites, fish, flying reptiles, whales, and humans? Probably not, but we don't know for sure. Extinction eliminates promising lineages—often early in the adaptive process—but this creates space for evolutionary innovations. Therefore, in our world at least, extinction continually provides new opportunities for different organisms that can explore new habitats and modes of life. This process "keeps the pot boiling" and may be necessary to achieve the variety of life forms, past and present.

The foregoing suggests that extinction may be a necessary ingredient in evolution, but the case is by no means solid. We will return to the question in later chapters, where we will see that much depends on whether extinction is random or selective in its choice of victims.

SOURCES AND FURTHER READING

Cuppy, Will. 1983. *How to become extinct.* Chicago: University of Chicago Press. Reprint of 1941 classic.

Darwin, Charles. 1859. *On the origin of species.* London: Murray. Available in many editions and reprints.

Erwin, T. L. 1988. The tropical forest canopy. In *Biodiversity* (reference below), 123–29. Summary of the pioneering research that led to the estimate that as many as forty million species may be living today.

Keyfitz, N. 1966. How many people have ever lived on earth? *Demography* 3:581–82. Source of estimate of the proportion of people still alive.

Raup, D. M. 1981. Extinction: Bad genes or bad luck? *Acta Geologica Hispanica* 16 (1–2):25–33.

Wilson, E. O., ed. 1988. *Biodiversity*. Washington, D.C.: National Academy Press. Collection of essays and research reports on problems of present and future extinctions; based on presentations at the National Forum on Biodiversity, held in Washington, D.C., in September 1986.

CHAPTER 2

A BRIEF HISTORY OF LIFE

My review of the history of life will be selective because the complete job, in a 5,000-word chapter, would require summarizing 700,000 years with each word. I will cover a few high spots relevant to extinction, as well as some aspects of the fossil record useful as background to the discussions that follow.

ORIGIN OF LIFE

In the beginning, there were bacteria. The earliest record of life on earth is from rocks in Australia, 3.5 billion years old (abbreviated 3.5 ga BP, for gigayears before present), only half a billion years younger than the earth's oldest rocks. These fossils are of one-celled, anaerobic, nonphoto-synthesizing bacteria; their cells lack nuclei and several other features of more advanced forms of life. Though labeled primitive, these organisms have been tremendously success-ful and still flourish in a variety of environments today.

The Australian fossils are assumed to be nearly the first life on earth. It is further assumed that life originated from nonlife as a result of spontaneous chemical reactions—also on earth. These assumptions (presumptions might be a better word) cannot be proven, and alternative theories abound. For example, it is not impossible that early chemical steps in the origin of life occurred elsewhere and came to earth from space. For most people who work on the problem, however, earthly origination of life is the simplest alternative. Re-search on the origin of life is an active but small field. Some of it is purely theoretical, some involves laboratory study of the plausibility of life's having originated on the early earth, and some entails searches for complex organic molecules in space.

Another nearly universal assumption is that all subse-quent life descended from the original life form through a continuous chain of ancestor-descendant pairs. This assump-tion looks good because all living organisms share biochem-ical traits. It is conceivable, of course, that life originated more than once on the early earth but that all except one life form died out early, leaving a single lineage as the ancestor of life as we know it. If this did happen, it was the first important species extinction.

Several years ago, I worked with Jim Valentine, a paleobiologist at the University of California (Berkeley), to test the idea that life might have originated more than once. We used some methods from the Gambler's Ruin problem

23

(coming up in the next chapter) to ask, What are the chances, if there had been many independent origins of life, that all but one of them would have died out? Our analysis showed that even if life had originated ten times, the likelihood is that, merely by chance, only one of the ten would have survived. With more than ten originations, it becomes likely there would be living descendants of at least two of the original life forms. If there were, indeed, several origins of life, with all but one attempt dying out, we are left with a bad genes–bad luck question: Did the best life form win out in a struggle for existence, or was the surviving lineage merely lucky?

COMPLEX LIFE

The long interval of geologic time between the oldest bacteria (at 3.5 ga BP) and the appearance of complex, multicelled organisms (at about 0.6 ga BP) is known as the Precambrian, meaning before the Cambrian period. Throughout the Precambrian, the fossil record is sparse, but it does show some evolutionary changes. For example, the first hard evidence of photosynthesis appears about 2 ga BP, and the eukaryotes (organisms with nucleated cells) appear about 1.9 ga BP.

From a human perspective, the Precambrian was a long interval of sluggish change, with global biodiversity dominated by a few anatomically simple organisms. During this time, however, major changes took place in the chemistry of the atmosphere and the ability of life to utilize its environment, changes probably vital to all subsequent evolution. Perhaps most important was the development of an oxygen atmosphere, a product of early plant life, which, in turn, made oxygen-breathing animals possible. Thus, our oxygen atmosphere was both a cause and an effect of the diversification of life.

You will note how many times I am using words like *probably* and *perhaps;* the study of early life and of conditions on the early earth is dicey. It calls for good guesswork and occasional speculation—but not more so than in many other areas of science.

About 600 million years ago (600 ma or 0.6 ga BP), all hell broke loose in organic evolution. The rock record suddenly contains abundant remains of complex and diverse organisms. One of the oldest assemblages is called the Ediacara fauna, discovered in 1946 by R. C. Sprigg, an Australian government mining geologist.

There is an irony here, for no professional paleontologist would have looked for fossils where Sprigg found them— in pure quartz sandstones that rarely contain fossil remains. Moreover, the rocks were known to be older than Cambrian assemblages of trilobites and other common fossils. Sprigg's job in Australia was to explore old lead mines, but he had a strong amateur interest in fossil collecting and kept his eyes peeled even in rocks where no paleontologist would bother to look.

The Ediacara fauna is now known worldwide. Its fossils are strange, soft-bodied, aquatic organisms. Some may belong to evolutionary groups living today, but most are enigmatic. One popular view among paleontologists is that Ediacara represents a major evolutionary branch that was killed off—a false start. In this regard, Ediacara is comparable to the somewhat younger Burgess Shale (Middle Cambrian of British Columbia), which Stephen Jay Gould has described and interpreted so elegantly in his book *Wonderful Life*.

Whatever the origin and fate of the Ediacara organisms, they were complex animals that occupied many regions during the latest Precambrian. Their occurrences are so close to the same age that the duration of the Ediacara reign is difficult to estimate: it was probably very short.

The start of the Cambrian period, about 570 ma BP, is marked by the appearance of much higher diversity. From then to the present day, most rocks that can contain fossils do.

Why, after such a long period of sluggish evolution, did life on earth suddenly diversify, a change so dramatic that it is often referred to as the Cambrian explosion? One theory holds that something happened in the physical environment—a change in composition of oceans or atmosphere perhaps—that stimulated development of a wider variety of organisms. Perhaps a sudden increase in availability of calcium carbonate in the oceans encouraged evolution of organisms that use calcium carbonate for hard skeletons and shells.

Or the diversification could have had a biological cause: the appearance of organisms that grazed on shallow marine areas covered with simple algal communities, thereby encouraging diversity. This idea comes from a well-known ecological principle called *cropping,* well argued by Steven Stanley of Case Western Reserve University. The presence of a consuming species, such as a grazer or a carnivore, stimulates species diversity in the area being cropped.

A rather different but appealing explanation of the Cambrian explosion likens this phase of evolution to a disease epidemic. Many disease organisms persist at low levels for many years and then, without apparent cause, expand to epidemic levels. Growth of any disease is exponential (like compound interest): the more disease organisms there are, the larger the number added (by reproduction) in a short time. When there are only a few individuals, population growth is not dramatic, but as populations expand, more and more individuals are added in each reproductive cycle and the disease becomes epidemic.

Evolution is indeed like a disease if one thinks of speciation as analogous to reproduction of the disease organism and extinction as analogous to its death. As long as the rate of speciation exceeds the rate of species extinction, the number of species (biodiversity) must increase exponentially. The more species there are, the more opportunities for further speciation. The long Precambrian, with its sluggish evolutionary expansion, may thus be analogous to a disease organism that has not yet reached the steep part of its exponential growth curve. If the analogy is correct, it is futile to search for some special event—physical or biological—that triggered the Cambrian explosion.

THE QUALITY OF THE FOSSIL RECORD

The fossil record is awful and superb at the same time. On the one hand, only a minuscule fraction of past life has been fossilized (and found by paleontologists). On the other hand, we have tens of millions of excellently preserved fossils to work with. About 250,000 species have been described, named, and located reasonably well in space and time. So, although the sample of past life is a small percentage of the whole, it is large enough to provide a lot of information.

There is another problem. The quality of the fossil sample varies enormously from organism to organism and from one physical environment to another. In general, aquatic organisms are more likely to be preserved than those on land because lakes and oceans are sites of sediment deposition. Animals with hard, mineralized skeletons fossilize more readily than soft-bodied organisms. Thus, the record of marine shellfish is far better than that of terrestrial insects.

A curious aspect of fossilization is that a plant or animal is most likely to be preserved if it is removed from the environment where it lived. The natural environments of most species are biologically active and support many scavenging organisms, including decay bacteria. If a plant or animal dies in that environment, its remains are soon consumed by scavengers. But if the dead body is moved quickly to a biologically inactive setting, preservation potential is enhanced. Our best fossil localities were formed in this way, including the La Brea tar pits, where numerous Pleistocene animals were caught in liquid tar. In a few cases, land animals were suffocated by showers of volcanic ash. Freak occurrences of this kind provide our most precious windows on the past.

600 MILLION YEARS OF FUSSING

By the end of the Cambrian period, global biology had developed elaborate and diverse communities, at least in the oceans. Dry land existed but was not yet occupied: no trees, no insects, and no flying organisms. Jacques Cousteau would have found it easy to make interesting and exciting television shows in this period, although he would have had to do without sharks and most fish. Still, there were tropical reefs to film, as well as some interesting swimming and bottom-dwelling animals. Cousteau would probably have concentrated on the trilobites, because of their large size and great diversity.

The 600 million years from Ediacara to the present, known as the Phanerozoic, provide an ample fossil record and most of our knowledge of evolution (and extinction). The Phanerozoic is often taught in school as a series of discrete intervals—the Age of Fishes, the Age of Reptiles, and so on—leading to the Age of Mammals and us. However, as Stephen Jay Gould has shown convincingly, the Phanerozoic was not really like that. It was not a succession of global dynasties dominated by organisms of ever-increasing sophistication.

True, the Phanerozoic saw a number of important additions to the biological repertoire. Many of these survive today and are still evolving. Dry land was invaded by plants in the Devonian period (for the time scale, see Figure 2-1), and highly evolved insect flight followed almost immediately. By the Carboniferous period, tropical rain forests were well developed (when conditions were right) and land-dwelling vertebrate animals came soon afterward. Some studies of fossil assemblages from Carboniferous rain forests suggest that insect diversity was locally as great as it is today.

From the Permian period onward, small and large vertebrate animals were abundant on land. It is often said that the large reptiles dominated the world, both on land and in the oceans, for long spans in the Jurassic and Cretaceous periods.

29

But this is a gross exaggeration. To be sure, some reptiles were the largest animals then living—big dinosaurs on land and ichthyosaurs and mosasaurs in the oceans. But in terms of the global biomass, they were minor players, never having many species or large populations compared with the millions of smaller organisms. Quite likely, no more than fifty dinosaur species lived at any one time. By contrast, more than five times that number of squirrel species are living today.

Following the extinction of dinosaurs and large marine reptiles at the end of the Cretaceous period, mammals diversified rapidly. This led in time to *Homo sapiens,* our own species.

Because there were many notable innovations in evolution during the Phanerozoic, including those just mentioned, some paleontologists have viewed it as a time of orderly progression from simple to complex, from primitive to advanced, and from small to large organisms. But such generalizations have not survived close scrutiny. In fact, the evolution of life during the Phanerozoic was dominated by backing and filling and a lot of general fussing. Major evolutionary groups appeared, flourished for a while, and then died out; the organisms replacing them were dif-

FIGURE 2-1. Standard geologic time scale for the younger part of earth history, showing eras, periods (Cambrian through Tertiary), and epochs of the Tertiary (Paleocene, Eocene, and so on). (The Pliocene epoch follows the Miocene but was too short to be labeled at this scale.) The classification of time is based primarily on fossils. The scale on the left is the current calibration (from Harland et al., 1990) of the fossil-based chronology.

GEOLOGIC TIME SCALE

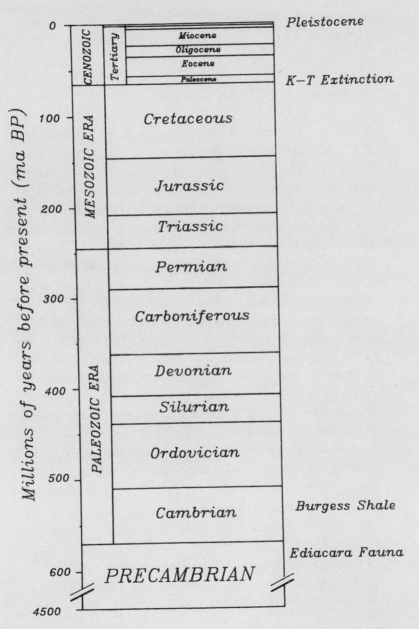

31

ferent but not demonstratively more complex, more advanced, or larger.

A STOCK MARKET ANALOGY

One gets the flavor of Phanerozoic history by comparing it to stock market tables over several decades. In the 1920s, the New York Stock Exchange used almost the same format as today—alphabetical listings of company names and indications of price, earnings, and so on. The total number of companies was somewhat lower, just as biodiversity in the Devonian was somewhat lower than it is today. Through the years, companies come into and go from stock exchange listings. Once extinct, companies do not return. Also, certain industries wax and wane. In the 1920s and 1930s, railroad stocks were numerous, whereas airlines, though present, were of negligible importance. Occasionally, new industries appeared and, if they lasted, went through periods of expansion and consolidation. All of this can be tracked by counting numbers of companies listed for each industry.

In any week, prices of stocks fluctuate chaotically. Sometimes nearly all stocks move in the same direction, whereas at other times some companies go one way and others go the opposite way, apparently independent of each other. The average price of all stocks at week's end is a consequence of numerous internal and external factors that influence prices. When the price of a stock goes to zero, the company is extinct. Above all, stock prices, as well as the composition of the entire market, are virtually unpredictable from week to week or decade to decade. And so it was with biological evolution in the Phanerozoic.

Although stock market history is a surprisingly good model for Phanerozoic evolution, the analogy breaks down in some regards. For example, the recent spate of corporate mergers is analogous to the biological process of hybridization: evolutionary lineages merge. This is not, as far as we know, common in evolution.

When stock market reports spanning fifty or seventy-five years are compared, one can see a definite, though irregular, shift toward companies with a more modern flavor. Plastics and aerospace companies thrive, and some well-known names like Xerox and Apple Computer appear. Inevitably, the list gradually approaches that of the present day. The same thing happens in the fossil record of the Phanerozoic. With each change in fauna and flora, the global biota becomes a little more contemporary. Because we have a vantage point at one end of the time sequence, the changes give an impression of a directed sequence leading to where we are. In other words, progress. But this impression would exist even if the evolutionary sequence were totally chaotic.

Because the perception of trends and patterns depends so much on one's vantage point, it is difficult to view the evolutionary record objectively. This is especially true when we deal with evolution of land-dwelling vertebrate animals: amphibians, reptiles, birds, and mammals. It is almost impossible to escape a feeling that the human species is the culmination of an upward progression—whatever "upward" may mean. This notion of progress implies that mammals are somehow better organisms than reptiles or amphibians and that humans are somehow better than other mammals. This, in turn, implies that extinctions in the past were due to the victims' deficiencies—in other words, to bad genes.

Some vignettes from Phanerozoic life will illustrate both the elegance and the confusion of the evolutionary record.

TRILOBITE EYES

Visual perception evolved independently many times in the animal world. In some cases, vision is limited to mere light-sensitive tissues, which, though valuable to the animal, hardly qualify as eyes. Simple ability to distinguish light from dark is found widely among sea urchins, starfish, and many other invertebrates. True eyes developed in such different groups as insects, mollusks, birds, and mammals. Although these groups have a common ancestor if one goes back far enough, their eyes are independent evolutionary inventions.

The trilobites of the Paleozoic era (570–245 ma BP) had compound eyes similar to those of modern crabs, insects, and other arthropods. In this case, the similarity is probably a result of common ancestry, although the fossil record is not available to prove this. The compound eye consists of many separate elements, with separate lenses, all working together to form an image. Occasionally, trilobite fossils are good enough to show the lens system of the eye in an almost unaltered state.

Several years ago, the trilobite lens system attracted a University of Chicago physicist and avid fossil collector, Riccardo Levi-Setti. Working in collaboration with a trilobite specialist, Euan Clarkson at the University of Edinburgh, Levi-Setti made some astonishing observations. In well-preserved fossil specimens, each element of the trilobite eye has two lenses, one above the other. Viewed from above, the interface between the lenses has a central depression with a rounded rim.

The two-lens nature of the trilobite eye is common in modern optical design and is called a doublet. But the shape of the upper lens is unlike any now in use either in nature or in man-made optics. Levi-Setti, with training and experience in optics, was able to recognize, however, that the shape of the upper lens of the trilobite eye is identical to designs independently published in the seventeenth century by Huygens and by Descartes. This lens shape was devised to minimize spherical aberration. The Huygens and Descartes designs were apparently never used, because other lenses were available to serve the same purpose.

The lower lens was the trilobites' idea. Levi-Setti was able to show that the doublet is necessary to avoid spherical aberration under water—something the seventeenth-century designers were not concerned with.

My point is that, even early in the Phanerozoic, organisms had evolved highly sophisticated systems—in this case, systems that in human terms would require a highly trained and imaginative optical engineer. Were trilobite eyes more effective than those of modern crabs or shrimps? We cannot answer this, because we cannot observe living trilobites. We can say only that there is no evidence that the eyes of the modern crab are better.

TROPICAL REEFS

Tropical oceans today are rimmed with massive and beautiful coral reefs, each a highly diverse and complex community of plants and animals. Cresting near the water surface and resistant to most wave action, reefs often create protected lagoons that support other diverse communities. The strong framework of most modern reefs is constructed of the skeletons of colonial coral animals of the order Scleractinia. Much of today's marine biodiversity is tied to tropical reefs.

Reefs are confined mostly to tropical oceans. This is due in part to climate (temperature) and in part to the high angle of the sun's rays in low latitudes—the corals being dependent on vigorous photosynthesis of symbiotic algae. Over the past few tens of millions of years, reefs have fluctuated toward and away from the equator as conditions have changed.

Deeper in the geologic past, tropical oceans have sometimes had well-developed reefs and sometimes not. Figure 2-2 summarizes this history, showing times when reefs were well developed (as today) alternating with intervals of reef absence and with relatively long periods of incipient reefs. By "incipient," I mean reefs limited to a few localized occurrences and lacking a robust framework.

Alternation between reef and nonreef biotas has depended in part on changing geography and climate: continents have moved and weather patterns have changed. But the most important reason for the presence or absence of reefs has been biological: availability of organisms to build the framework. In Figure 2-2, all time intervals during

FIGURE 2-2. Summary of occurrences of tropical reefs through time. Note that fully developed reefs existed less than half the time. Intervals of a total absence of reefs invariably follow mass extinctions. (Adapted from Copper, 1988.)

Tropical Reefs through Time

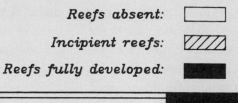

Reefs absent:

Incipient reefs:

Reefs fully developed:

Tertiary	
Cretaceous	
Jurassic	
Triassic	
Permian	
Carboniferous	
Devonian	
Silurian	
Ordovician	
Cambrian	

which reefs were absent follow immediately after major mass extinctions. Evidently, mass extinctions eliminated critical reef species, and, except for the most recent wipeout (end of Cretaceous), many millions of years were needed to re-evolve reef communities. Each time reefs developed again, their species composition differed from preextinction communities.

Today's corals (scleractinians) are a relatively recent evolutionary development: they do not appear in the fossil record until the Middle Triassic, about 240 ma BP. Yet reefs of the same basic construction and ecology are found back almost to the beginning of the Phanerozoic. Those reefs were built by an amazing variety of different organisms.

Many early reefs used calcareous algae as the main framework organism. Later, sponges were used; then a number of now extinct coral groups anatomically different from modern reef corals. Even a kind of clam, called a rudist, built reefs. Rudist reefs were especially prominent in Cretaceous seas but died out completely, at (or shortly before) the mass extinction that ended the Cretaceous. Rudists were highly unusual clams, often confused by students with corals because the rudists adopted some aspects of coral morphology.

The history of tropical reefs is typical of ecosystem history in general. It all seems rather aimless—a series of sudden shifts from one system to another, changes often driven by widespread extinction of incumbents.

FLYING REPTILES

How wonderful it would be to see reptiles with a wing span of fifty feet soaring over modern coastlines! Flight, like vision, has evolved many times. Also like vision, it has sometimes been elegant, as in the flying reptiles, and sometimes very simple, like the rather crude attempts of flying squirrels and flying fish today.

The flying reptiles, known collectively as pterosaurs, existed from about 200 ma BP to the end of the Cretaceous, 65 ma BP. The giant *Pteranodon* was considerably larger than any bird living today and larger than many airplanes. Its featherless wings were formed of large folds or extensions of skin suspended from greatly lengthened bones of fingers, one of each hand, rather like the bat's.

Whereas some larger pterosaurs probably depended mostly on soaring (gliding on air currents), they were capable of taking off from level surfaces and of powered flight. Extensive theoretical and experimental work has been done with pterosaur models, including wind tunnel experiments, and the results make clear that pterosaurs could have been competent flyers. But we have no way of knowing just how competent they were. These gigantic animals coexisted with birds for most of their time on earth. But we don't know how well birds were flying then.

Flying reptiles left no descendants. They were merely another successful group that appeared, flourished for a time, then disappeared. Lest I give the impression of the pterosaur interval as being short-lived, note that their time on earth was thirty times longer than ours, so far.

Human Evolution

Humans are latecomers in evolution, and their history is difficult to study because the fossil record is incredibly bad.

Most of our ancestors occupied upland regions where fossils are rarely preserved. Also, human history is short and populations were small. Extinction does not seem to have played a large role in our ancestry, although the fragmentary fossil record may be hiding a lot of information.

One important point must be made, however, about human evolution in the context of the history of life. We generally think of humanoid intelligence as being the most significant aspect of our species, and this is probably true. But high intelligence could have evolved at almost any time in the Phanerozoic and in almost any biological group—in reptiles, in fish, in insects, and even in trilobites. To suggest that insects could have developed intelligence like ours sounds a bit extreme; after all, they have small brains and are obviously dumb. But I know of no neurological or other reason why intelligence could not have evolved in insects, accompanied by changes in anatomy and embryology to accommodate it. By the same token, intelligence need not have evolved at all.

These reflections become important when one considers the likelihood of humanoid intelligence among extraterrestrial organisms, if such exist. Ongoing research projects such as NASA's SETI Program (Search for Extraterrestrial Intelligence) have taken this question seriously. A commonly held view is that any evolving biological system should be expected to pass through pretty much the same sequence of evolutionary stages—leading to intelligence, a technological civilization, and the invention of radio communication. Although proponents of this view do not insist that extraterrestrials be humanoid in appearance, they expect the behavior and intelligence of the ETs to be similar to ours. This viewpoint has been unpopular with paleontologists and biologists because the Phanerozoic record shows no evidence of such predictability or consistency. Whereas I agree with this objection, I remain a strong supporter of the SETI effort. Only by discovering biological systems elsewhere in space will we really have the means to know whether our own biological system has predictable patterns not yet recognized.

LIVING FOSSILS

We have all heard of species that have survived unchanged for hundreds of millions of years. Common examples include the cockroach, horseshoe crab, shark, coelacanth fish, ginkgo, and horsetail. The usual presumption is that these species are winners in the evolutionary struggle, having found ideal niches and thus become immune to extinction. They are also presumed to have survived because they are generalized types that can live under harsh conditions and survive on almost any food. The shark is perhaps the quintessential living fossil. Sharks are not very bright but are strong, hard to kill, and able to eat anything, dead or alive. Sharks also have a distinctly primitive appearance.

I suspect that nearly everything I have written in the previous paragraph is bunk! In the first place, in none of the examples is the living species the same as the fossil species. There are hundreds of living shark species, as different as nurse sharks and hammerheads, and all are distinguishable anatomically from the ancestral sharks. Fossil horseshoe crabs from the Jurassic look much like their living counterparts, but this is largely a subjective impression stemming from the fact that horseshoe crabs, then and now, are very different from other common crabs. In detail, the Jurassic types are not confused with those living today.

The history of life has indeed experienced a wide range of rates and amounts of evolutionary change. Most species we cite as living fossils, like the coelacanth, are probably at the slow end of this range, but there is no evidence that this constitutes a discrete kind of evolution. More important, there is no evidence that organisms have ever evolved an immunity to extinction.

In this chapter, I have emphasized the lack of predictability in the sequence of evolutionary events that constitutes the history of life. At no point in the fossil record can we look at a particular event and say, "Of course, it had to happen that way." If flying reptiles had not evolved, no anatomist or physiologist would question their absence. By the same token, we do not know whether the biologically possible body plans or ways of life of organisms have been exhausted. If we could imagine all possible designs of organisms, we are unable (so far) to say whether most or only a small fraction of the designs have been tried. Organisms that use wheels or wind-powered sails for locomotion are virtually absent (although this depends a bit on how one defines wheels and sails). Are wheels and sails impractical or impossible in a living organism, or might they be feasible but not yet discovered by evolution? This level of uncertainty in evolutionary biology frustrates scientists in other disciplines who are accustomed to greater consistency and predictability.

Having insisted that the course of evolution over the last 3.5 billion years could not have been predicted, I must add a caveat. The appearance of disorder may only reflect our ignorance: there may be clear patterns in evolution that we have not detected even though the data are before us. Also, if we ever discover life in outer space, we will finally be able to judge which, if any, characteristics of life on earth are "normal" or inevitable.

SOURCES AND FURTHER READING

Clarkson, E. N. K., and R. Levi-Setti, 1975. Trilobite eyes and the optics of Descartes and Huygens. *Nature* 254:663–67. Source for discussion of trilobite optics.

Cloud, P. E. 1988. *Oasis in space: Earth history from the beginning.* New York: W. W. Norton. Well written for a general audience; emphasizes the early earth and the Precambrian record.

Copper, P. 1988. Ecological succession in Phanerozoic reef ecosystems: Is it real? *Palaios* 3:136–52. Research article used as the basis for Figure 2-2.

Gould, Stephen Jay. 1989. *Wonderful life.* New York: W. W. Norton. Penetrating analysis of the Burgess Shale fossils and the meaning of extinction in evolution.

Harland, W. B., et al. 1990. *A geologic time scale 1989.* Cambridge: Cambridge University Press. Standard for the estimates of geologic time used in this book.

Margulis, L. 1988. The ancient microcosm of planet earth. In *Origins and extinctions,* ed. D. E. Osterbrock and P. H. Raven. New Haven: Yale University Press, 83–107. Excellent summary of Precambrian life.

Nitecki, M. H., ed. 1988. *Evolutionary Progress.* Chicago: University of Chicago Press. A collection of essays on the science and philosophy of defining progress in the evolution of life.

Padian, K. 1988. The flight of pterosaurs. *Natural History,* December, 58–65.

Raup, D. M., and S. M. Stanley. 1978. *Principles of paleontology.* 2d ed. San Francisco: W. H. Freeman. College text.

Raup, D. M., and J. W. Valentine. 1983. Multiple origins of life. *Proceedings of the National Academy of Sciences* 80:2981–84. Research article exploring the possibility that life originated more than once.

Schopf, J. W., ed. 1983. *Earth's earliest biosphere: Its origin and evolution.* Princeton: Princeton University Press. Detailed and comprehensive collection of articles on Precambrian environments and early life.

Schopf, J. W., and C. Klein, eds. 1991. *The proterozoic biosphere.* Cambridge: Cambridge University Press. Forthcoming. An even more comprehensive treatment of Precambrian life.

Stanley, S. M. 1973. An ecological theory for the sudden origin of multicellular life in the late Precambrian. *Proceedings of the National Academy of Sciences* 70:1486–89. Research article outlining the case for cropping as the cause of the Cambrian Explosion.

Stanley, S. M. 1989. *Earth and life through time.* 2d ed. New York: W. H. Freeman. A college text.

CHAPTER 3

GAMBLER'S RUIN
AND OTHER PROBLEMS

GAMBLING

Suppose you are a casino gambler fortunate enough to find an even-odds game—you and the house each have a fifty-fifty chance of winning on each play. This might be a roulette table with no zeros (green numbers), if such there be. You are playing only red or black on this wheel, the two colors occurring with equal frequency. You have entered the game with ten dollars and bet one of them on red. If the wheel comes up red, you win one dollar and now have eleven; if it comes up black, you lose one dollar and have nine. Continuing in this manner, your stake will fluctuate in steps of one dollar until one of three things occurs: (1) you go broke, (2) the casino goes broke (or asks you to leave

because you are winning too much), or (3) you run out of
time or otherwise decide to leave.

Figure 3-1 shows several possible outcomes of the casino
scenario. Though simulated with a random-number genera-
tor on a home computer, they could just as easily have been
created manually by flipping a coin or drawing randomly
from a deck of playing cards, using red and black. In each
case, our gambler ultimately went broke, but in one of them
(Game #3), he/she did pretty well for a while.

The Gambler's Ruin problem, just described in its simplest form, has been used for years by statisticians as a model
of kinds of probability. Inevitably, a specialized language
has grown up around Gambler's Ruin. For example, the
paths followed by the gambler's fortunes in Figure 3-1 are
called *random walks*. Once started, the random walk has no
tendency to return to a level previously occupied. If the
gambler starts with ten dollars, no force induces the path to
stay near ten or return to ten. The system has no memory.
Every good gambler knows this, of course: a long string of
failures does not change the odds for the future.

The horizontal base of each graph in Figure 3-1 denotes a
zero stake—the level at which the gambler has lost the
original stake and is broke. This is called an *absorbing boundary* because if the path reaches this level, there is no return;
the game is over. We could change this condition by stipu-
lating that when you go broke, the casino gives you a single
dollar to keep going. In that case, the bottom of the graph
becomes a *reflecting boundary*—you bounce back with at
least one upward step. I know of no casino that does this,
except by offering credit.

In Game #3 of Figure 3-1, the gambler's stake dropped
from 10 to 1 early in the play, then rose to 14, and ulti-

Gambler's Ruin

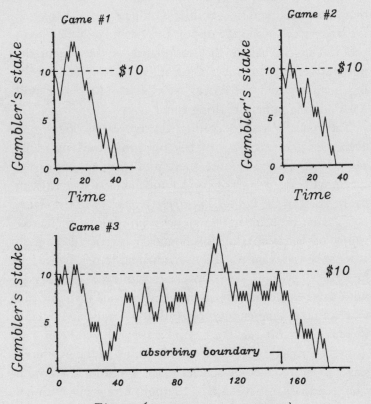

FIGURE 3-1.   Simulated gambling results in an even-odds game (fifty-fifty chance of winning on each play). The gambler's initial stake is $10, and each bet is $1. Thus, the stake fluctuates up and down randomly, in steps of $1. Gambler's Ruin occurs when the absorbing boundary—zero—is reached. Each game is like the fate of a genus that starts with ten species. The number of species goes up when a species branches to form another species and goes down when a species goes extinct.

mately dropped to 0 (the absorbing boundary). Suppose the gambler had started with nine instead of ten dollars and the play was otherwise identical. Zero would have been hit in that initial drop, and the gambler would never have enjoyed the later success—and the opportunity to leave with a small profit. This emphasizes the importance of the size of the initial stake: the higher it is, the farther it is from the absorbing boundary and the more likely that the gambler will remain in the game for a long time.

Theoretically, casinos could make a profit by offering an even-odds game as long as they put limits on how much players bet. This is because the typical gambler enters the casino with an amount of money much closer to zero than to the assets of the house. The *upper* boundary in the Gambler's Ruin problem is also an absorbing boundary (bankruptcy of the casino), but this boundary is generally so high as to be not relevant for the individual customer. Consider, however, a high roller who enters our even-odds casino with assets equal to precisely half those of the house. If this gambler plays long enough or is allowed large enough bets, there is a fifty-fifty chance of his breaking the house.

In the extinction context, we may think of the gambler's stake as the number of species in an evolutionary group. Our example's initial stake of 10 might be a genus with ten species living at some instant in the geologic past. We will use a time scale in millions of years instead of the gambler's time scale. For every interval of one million years, each species has a fifty-fifty chance of surviving to the start of the next million year interval; if it survives, it has a fifty-fifty chance of speciating to produce an additional species. What predictions can we make about the fate of the genus?

Several interesting predictions are possible. For example,

the number of species (diversity) will fluctuate up and down as in a random walk. Extinction of species lowers diversity; speciation increases diversity. As long as the chance of extinction is identical to that of speciation (fifty-fifty), a random walk will result.

Furthermore, *eventual extinction of the genus is inevitable.* This is somewhat counterintuitive, but it follows from the presence of the single absorbing boundary at zero species. As we have seen, a random walk is free to wander up and down indefinitely. If there is no *upper* absorbing boundary, the random walk is bound to hit the *lower* boundary eventually.

We could, of course, specify an upper absorbing boundary, analogous to the casino's total assets. In the context of global biology, the upper absorbing boundary would be all the spaces for organisms in the world. An evolutionary group, such as a genus, could "break the bank" by speciating so many times that no other genus could exist. All species in the world would belong to the same genus. This is as unlikely as an ordinary gambler's winning the whole casino. Thus, for all practical purposes, the ultimate extinction of the genus is assured.

Concepts of Randomness

What is the meaning of randomness in the natural world? The flip of a coin is said to be random. To most people, this means that the outcome is a matter of pure chance—without cause. But coin flipping is surely governed by cause and effect. Whether the coin lands with head- or tail-side up depends on an almost uncountable number of physical factors, including the side facing up at the start and the strength of the flip—and therefore the number of times the coin turns in the air. Also relevant must be the state of air currents (including wind) and perhaps also barometric pressure, to say nothing of the condition of the coin. (In the somewhat similar case of horseshoe tossing, by the way, good players become expert at controlling the number of turns of the shoe in the air.) Coin flipping is so complex that we *cannot* or *choose not to* monitor all determining factors.

Instead, we choose to assume that the complex of causes will combine to make the coin behave *as if* the process were random. Having made this choice, we can ignore all the wind currents and other details and adopt the statistical presumption that heads and tails are equally likely. This, in turn, makes available an arsenal of mathematical procedures for dealing with random events, enabling us to answer questions like, How often should tails be expected to come up twenty times in a row? The assumption of randomness is a clever dodge. By choosing to ignore cause and effect and all the complexity that goes along with it, we have a tractable phenomenon about which we can make interesting and useful predictions.

Most scientists and philosophers now agree that nothing is truly random in the natural world. The motion of molecules in a gas, the advance of a glacier, the formation of a hurricane, the occurrence of an earthquake, and the spread of an epidemic all have causes. In some cases, it is possible and worthwhile to investigate causes. This is certainly true for earthquakes and disease. But in other cases, we cannot or choose not to learn all there is to know. For example, by assuming randomness in the movement of molecules in a gas, we can derive the gas laws (Boyle's law and the like), which are critical in countless engineering applications.

Perhaps the best working definition of randomness in natural systems is the following: *Random events are events that are unpredictable except in terms of probability.* The 70 percent rain forecast is a case in point.

Using the Gambler's Ruin approach to extinction, we deal with the same sort of probabilities. We assume that there is some number of reasons, probably large, for a species to go extinct. We observe patterns of extinction in the fossil record that are indistinguishable, at some scale, from patterns that could have been produced by a purely random process. Regarding the fate of a species, this is tantamount to saying that if one stress or calamity doesn't get you, another will. Once we have made the assumption of random behavior, we are free to work with the patterns. This approach makes possible generalizations that would be out of reach of the more conventional case-by-case approach.

Gambling for Survival

In the evolutionary history of a genus or any other group of related species, there is a certain amount of randomness in the sense just defined. A complex of physical and biological factors determines how long each species in a genus will survive and whether species will branch to form new species. Species extinction weakens the future of the genus, and speciation protects it. It has been said that genera do not struggle for existence—they gamble.

By good luck alone, a genus (or any other group) may thrive for a considerable time, just as a gambler may have some dramatic successes. Furthermore, if a run of good luck has produced a lot of species—equivalent to a winning streak at the casino—the chance that the genus will go extinct in the next few million years is lessened. A large number of species thus provides the group with temporary protection from extinction.

There are more species of rodents today (about seventeen hundred) than of any other order of living mammals. Next in diversity are the bats, some nine hundred species strong. Thus, almost two-thirds of all living mammal species are rodents or bats. Could this be just luck? Did these two groups enjoy a run of speciation or a lucky avoidance of extinction early in their history? Or are they simply good at surviving and/or speciating for some definite biological reason?

One problem in deciding between these alternatives is that random processes have a wide range of outcomes; the position in that range is largely unpredictable. The question for the rodents or bats is as follows: Is the evolutionary history of the group within the expected limits of an even-odds game, or is the surplus of speciations beyond reasonable statistical expectation? If the latter is true, they must must be doing something right. Will Cuppy, in *How to Become Extinct,* thought he knew the answer. He wrote, "Bats are going to flop, too, and everybody knows it except the Bats themselves."

DIFFERING EXTINCTION AND SPECIATION RATES

In all of this, I have been assuming an equality of speciation and species extinction rates. Birth and death of species have been assigned precisely the same likelihood. How could this be realistic, especially since species birth and species death are such different phenomena?

There are two responses. First, the random-walk logic works perfectly well with *un*equal probabilities—only the mathematics changes slightly. To return to the casino analogy, the house normally establishes an imbalance in the odds so that its profit is favored. The random walk still works but with a bias slightly against the customer. Therefore, we can easily handle cases in evolution where the likelihoods of speciation and extinction are different. We may suspect, for example, that the rodents and bats have lived in situations where speciation is more likely than extinction. We must be wary, as was noted earlier, that the rodents and bats were merely lucky enough to beat the odds. Some casino gamblers do win and win consistently.

The second response to the question about the assumption of equality of rates is to note that the total number of speciation events in the history of life is approximately the same as the total number of extinctions. This follows from the 1,000:1 ratio between extinct and living species discussed in the first chapter. If 40,000,000,000 species were formed in the past and if 40,000,000 species are living today (keeping in mind that we do not know the exact numbers), then there have been 40,000,000,000 speciations and 39,960,000,000 species extinctions. Thus, the long-term average rates of speciation and species extinction have been about the same. Regardless of the reasons for this near-equality of rates, the numbers indicate that the even-odds model is not unreasonable.

SKEWED HISTOGRAMS

Paleontologists have done many computer simulations in order to gain a feel for the range of outcomes of random walks in biodiversity, with both equal and unequal rates of speciation and extinction. Some simulated groups expand to the point of swamping the computer's memory; other groups go extinct rather soon. Extinction of a group is most common when the group starts out small, just as a casino gambler is most likely to go broke quickly if he starts with a small stake, close to the absorbing boundary.

In evolution, a group such as a genus or family must, by definition, start with a single species. For the fledgling group to survive, the founding species must speciate before it goes extinct. Because new evolutionary groups start small, they usually don't last long. This, in turn, yields an important facet of the history of life: most groups of species have life spans shorter than the average of all groups. Figure 3-2 shows a histogram of life spans of fossil genera. It has a *skewed* shape, with many short durations and only a few long ones.

The skewed (asymmetrical) shape of variation is typical of important biological properties germane to the extinction question. These include

- numbers of species in a genus
- life spans of species
- numbers of individuals in a species
- geographic ranges of species

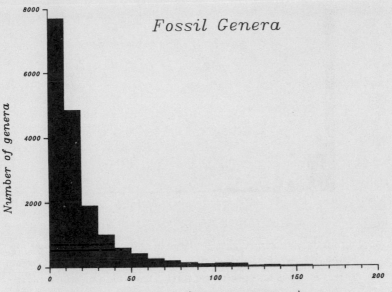

FIGURE 3-2. Histogram showing the distribution of geologic life spans of fossil genera. The mean duration of fossil genera is about twenty million years. The graph is highly skewed—many more genera have life spans less than the mean than have life spans greater than the mean. (Based on time ranges of 17,505 genera; tabulated by J. J. Sepkoski, Jr.)

In each category, the small "thing" is most abundant. Let me give another example. There are about 4,000 living species of mammals, grouped into about 1,000 genera. About half these genera have only a single species, and about 15 percent have only two species. The numbers drop off smoothly (see Figure 3-3), so there are only a few genera with more than 25 species. The most speciose

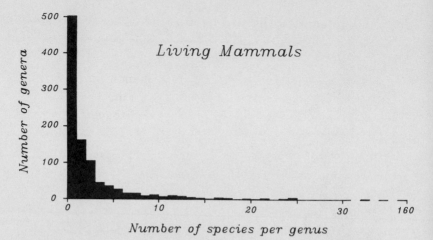

FIGURE 3-3. Skewed distribution of genus size among living mammals. About half the genera have only one species. Genera with as many as ten species are rare. (Based on data from Anderson and Jones, 1967.)

living mammal genus (a small insectivore) has about 160 species. The overall average is 4 species per genus (4,000/1,000), but because of the asymmetry, fully three-quarters of the genera have 1, 2, or 3 species and are thus below average.

Let me summarize a few of the foregoing points. For reasons that develop from both theory and observation—some depending on the Gambler's Ruin problem—we can make the following generalizations:

1. Most species and genera are short-lived (compared with the averages).
2. Most species have few individuals.

3. Most genera have few species.
4. Most species occupy small geographic areas.

Skewed variation is extremely common in nature. Strangely, however, most of us have been trained to believe that variation in natural phenomena is bell shaped, having just as many items above as below the average—whether we are talking about heights or weights of people, weather, or baseball averages. Nothing could be further from the truth.

Classic examples of skewed variation include incubation times of infectious diseases and life expectancies of cancer patients. In both, the majority of cases fall below the average, because the average is constructed by summing many short time intervals and a few long ones and then dividing by the total number of cases. It would be far better to use the *median* time—that time exceeded by half the individuals.

Of course, bell-shaped curves (called normal or Gaussian distributions by mathematicians) do sometimes occur in nature. It is only that other shapes are more common. Statisticians wrestle with this problem because many of the best statistical tests are designed for the bell-shaped curve. Often they avoid the problem by transforming the raw data—that is, distorting the scale of measurement so that they can treat the results *as if* they had a bell-shaped distribution. One such transformation that sometimes works is to convert all the measurements to their logarithms (or even square roots). If the transformed numbers have a bell-shaped distribution, the analyst can proceed with tests that assume this shape.

OTHER MODELS

The Gambler's Ruin problem has led us to generalizations about species that are relevant to the extinction problem. However, many of the patterns, especially the skewed distributions, can also be approximated by processes having nothing to do with gambling or biology.

Suppose you take a stick that is 100 inches long and break it at 25 random points—not favoring the middle or any other part. When you are done, you will have 26 short sticks. Now, measure and count the short sticks and construct a histogram. The shape of variation in stick length will look very much like those I have shown for species and genera: a hump or spike to the left and a long tail extending to the right. Figure 3-4 shows the results of a computer simulation.

Variation in population sizes of cities in the United States shows a pattern like this, as do many other things we can measure or count. The so-called broken stick model is one of several that have been applied to these patterns, and many attempts have been made to find out which model makes most sense or fits the observations best. For the purposes of this book, the important thing is that many of the distributions are skewed. They are not even close to the symmetrical, bell-shaped curve that we have all heard about.

One lesson about extinction to be learned from this is that some plants and animals are much more likely, a priori, to go extinct than others. The majority of species living today have small populations and live in restricted geo-

Broken Stick Model

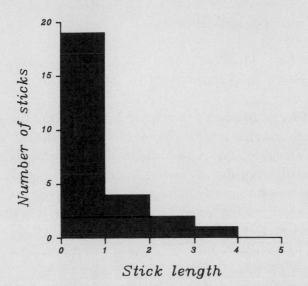

FIGURE 3-4. Results of a simulation wherein a stick is broken at 25 points selected at random. The histogram shows the size distribution of the short sticks. The extreme skewing recalls shapes commonly seen in histograms of natural phenomena.

graphic areas. These are the ones we rarely see. The abundant and widespread species are commonly seen but are surprisingly few in number. For this reason, it is possible to write useful field guides to mammals and insects in volumes of manageable size. It stands to reason that when things get environmentally tough, either biologically or physically, the many rare species are the most vulnerable to extinction. So, when we say that a given extinction event eliminated 40 or 80 percent of biodiversity, we should also say which 40 or 80 percent. The significance of the event will depend heavily on whether the victims were abundant, cosmopolitan species or local endemics.

A Note on Extinction of Surnames

Thomas Malthus, a British economist of the late eighteenth and early nineteenth centuries, is remembered for his ideas on population growth and its societal effects. His argument that populations always increase more rapidly than food supplies strongly influenced Charles Darwin in his development of the theory of natural selection. It is strange, therefore, that Malthus's most famous work, his *Essay on the Principle of Population,* should provide an important twist on the extinction problem.

A small section of the *Essay* was devoted to population figures for the town of Bern, in Switzerland. Malthus compared lists of the names of prominent families—meaning families registered as members of the bourgeoisie, such as merchants, and artisans—over a 200-year period from 1583 to 1783. To Malthus's apparent astonishment, fully three-quarters of the families listed at the start of the period were gone by the end (even though the total population size was stable). All people with those surnames died out or left Bern. Malthus gave no explanation for the observation, and the case went largely unnoticed in a piece devoted to the growth rather than the death of populations.

Through the nineteenth century, high family extinction rates were noted in other groups for which good records were available, especially European royal families and the

English peerage. The whole thing was counterintuitive. Everyone was aware of enormous families with names that could be traced back for centuries. But the statistics were firm.

High levels of surname extinction were interpreted by most observers to betray some fundamental weakness or debility in upper-class life. The bourgeoisie and royal families were somehow less likely to live long or spawn enough children to keep their families going. This interpretation was based on the tacit assumption that data from the proletariat would show more stability. For many years, nobody bothered to check.

A classic mathematical study published in 1875 by Francis Galton and H. W. Watson showed, however, that the original observations of Malthus were exactly what should be expected. And when broad-based census data for whole populations were analyzed, Galton and Watson's conclusion was confirmed.

So, most family names in human communities have a surprisingly short life. A new family name usually starts out with one or a very few individuals, and whether the name persists for many generations depends on the vagaries of chance: how many male (name-bearing) children are born and whether they survive to sire children. The size of any family fluctuates like a random walk. If the overall birthrate is high enough to cause population growth, the random walk is slightly biased toward persistence of family names, but this bias is not strong enough to avoid Gambler's Ruin, especially because new families start out small. Therefore, the very large families Americans are familiar with—Smiths and Johnsons—are a rarity. The success of these families is perhaps analogous to that of the rodents and bats.

As a final comment on human surnames, what happens in a human population over thousands of years? Because of the extinction of names, the population must become concentrated among fewer and fewer families, with the end result (given infinite time) of a community with only one family. Because surnames serve primarily to distinguish or label family lineages, there must be some optimal number in any community. When names become too few because of extinction, I imagine there is a strong incentive for families to split and new names to appear. This is an evolutionary splitting but one that does not seem to have a clear analogue in the biological world. Or does it?

SOURCES AND FURTHER READING

Anderson, S., and J. K. Jones. 1967. *Recent mammals of the world.* New York: Ronald Press. Source for Figure 3-3. The word *Recent* in the title is jargon for "living" (as opposed to "extinct"). An updated volume was published in 1984 under the title *Orders and families of recent mammals of the world* (New York: John Wiley), but the statistics have not changed appreciably.

Dubins, L. E., and L. J. Savage. 1976. *Inequalities for stochastic processes—How to gamble if you must.* New York: Dover Publications. A heavy mathematical treatment based on a casino gambling model; includes a prescription for ensuring that a gambler will end a winner in 90 percent of casino outings—while losing all the profits the other 10 percent of the time.

Galton, F., and H. W. Watson. 1875. On the probability of the extinction of families. *Journal of the Anthropological Society of*

London 4:138–44. A classic study of the extinction of family names.

Malthus, T. R. 1826. *An essay on the principle of population.* 6th ed. London: John Murray. Pp. 352–53 of vol. 1 contain the data on extinction of families in Switzerland.

CHAPTER **4**

MASS EXTINCTIONS

Mass extinction is box office, a darling of the popular press, the subject of cover stories and television documentaries, many books, even a rock song. *Discover* magazine devoted its October 1989 issue to the topic "A Decade of Science: The Eight Big Ideas of the Eighties." Idea no. 5, described in an essay by Stephen Jay Gould titled "An Asteroid to Die for," concerned extinction. At the end of 1989, the Associated Press designated mass extinction one of the "10 Top Scientific Advances of the Past Decade." Everybody has weighed in, from the *Economist* to *National Geographic*.

There are several reasons for the popular excitement about extinction. An obvious one, and certainly the most important, is the controversial research article published in 1980 by L. W. Alvarez, W. Alvarez, F. Asaro, and H. V.

Michel. This article proposed that a giant comet or asteroid collided with earth 65 million years ago, triggering one of the largest mass extinctions in the history of life. The fact that the extinction killed the dinosaurs added to popular interest.

Then, in 1984, Jack Sepkoski and I published the claim that the last several pulses of extinction were clocklike in their spacing, coming approximately every 26 million years. This led astronomers to propose solar system or galactic explanations, the most notable being the theory that our sun has a small companion star (dubbed Nemesis) that disturbs comet orbits every 26 million years, causing showers of comets on earth. This combination of proposals touched off yet more controversy.

These scientific events would have sufficed to raise excitement about mass extinction, but additional explanations have been suggested, including the notion that doomsday scenarios have natural appeal in our culture. The big mass extinctions were truly devastating events and, if caused by collisions with comets, instantaneous and dramatic. Could it happen again? And if so, when?

It has also been suggested that popular fears of global war and nuclear winter have sensitized all of us to talk of global disaster. If that's true, global warming and the greenhouse effect are probably part of the mix.

When a paleontologist is asked how many mass extinctions there have been, the invariable answer is five: one each in the Ordovician, Devonian, Permian, Triassic, and Cretaceous periods—events known as the *Big Five*. When asked what went on in between, he or she usually replies that there was a continuous, low level extinction called "background" with, perhaps, a few pulses above background but not large enough to be called mass extinctions.

Figure 4-1 attempts to show the timing of extinctions of varying size. The long, open arrows are the Big Five; the one at the end of the Permian (245 ma BP), the biggest. The shorter arrows indicate smaller events, with the arrows varying in length roughly according to intensity of extinction.

This raises several questions. How is extinction measured? Are intervals of high extinction sufficiently short-lived to merit the word *event?* Are there fundamental differences between large and small extinctions—other than size—as the terminology we use implies? Before tackling these questions, I will give a brief description of one big mass extinction.

THE K–T MASS EXTINCTION

The best-documented of the Big Five came at the end of the Cretaceous. It is often called the K–T—referring to the boundary between the Cretaceous (abbreviated K, to avoid confusion with the Carboniferous and Cambrian periods) and the next younger, or Tertiary (T), period. Because it is the most recent of the Big Five, its rocks and fossils are the best preserved. Also, sediments from the Cretaceous are widely distributed because it was a time when continents were flooded by shallow seas, leaving a good marine record on the present land surface.

Virtually all plant and animal groups—on land and in the sea—lost species and genera at or near the end of Creta-

Major Extinction events

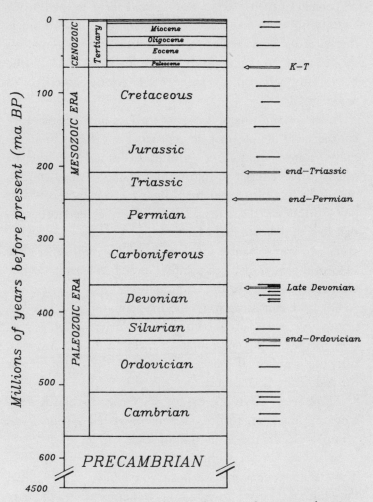

FIGURE 4-1. Geologic time scale showing the principal extinction events of the Phanerozoic. The arrow lengths are roughly proportional to the intensity of extinction. The labeled events are the Big Five mass extinctions. (Adapted from a similar diagram constructed by Sepkoski, 1986.)

67

ceous time. Marine animals suffered the total extinction of 38 percent of their genera; among land animals, the hit was slightly higher. These are big numbers when one considers that in order for a genus to die out, all individuals in all its species must go. Land plants appear to have done a little better, although their fossil record is not good enough for us to be sure.

In the oceans, major losses of species and genera were concentrated among marine reptiles, bony fishes, sponges, snails, clams, ammonites (mollusks distantly related to squids), sea urchins, and foraminifera (single-celled animals usually having a hard skeleton). But no group escaped. Most noteworthy are the large groups—families and orders— that lost all their species. Marine reptiles (plesiosaurs, mosasaurs, and ichthyosaurs), ammonites, and several other once successful groups died out totally. Some had been in decline well before the end of the Cretaceous; others died abruptly.

On land, dinosaurs were the most obvious victims, but heavy losses were sustained by a wide variety of other reptiles, mammals, and amphibians. In western North America, fully one-third of all genera of mammals died out at or near the end of the Cretaceous.

There were survivors, of course. Crocodiles, alligators, frogs, salamanders, turtles, and mammals all survived—as groups—despite the extinction of some species.

In a way, lists of victims and survivors are misleading. The statement that mammals survived hides the fact that they sustained heavy losses. Too much emphasis on the fates of major taxonomic groups has led to oversimplification in the search for causes of mass extinction—posing false questions like "What could kill dinosaurs but not affect mammals?"

More important, the lists hide more dramatic effects of the extinction—huge die-offs of a few, very abundant species. For example, many K–T boundary deposits show a sudden shift in land vegetation recorded by a drop in angiosperm pollen (flowering plants) and a jump in fern spores. This is known as the fern spike. In a few millimeters of the sedimentary sequence, fern spore content goes from 25 percent to 99 percent. This change is reminiscent of sudden shifts of vegetation often seen after a forest fire today—a lush forest is replaced by an opportunistic flora dominated by ferns.

Following the fern spike, flowering plants came back with relatively few species missing. If one were to consider only counts of species and genera, the trauma of the event would be overlooked. And this was Digby McLaren's argument on the significance of mass killings in the past (see Chapter 1).

Another mass killing occurred in the oceans among near-surface foraminifera ("planktonic forams"). These small animals had only a few species, but they were hugely abundant, their skeletons often dominating sedimentary deposits of the time. At the K–T boundary, most planktonic forams were killed off so completely that the overlying sediments are strikingly different in color and general appearance. Farther up in the Tertiary sequence, the planktonic forams come back, with several new species descended, apparently, from a single surviving Cretaceous species. Because few species were involved, the planktonic forams do not add much to global statistics of the event. But in biomass loss, they were important.

Unfortunately, the fossil record rarely allows us to count numbers of individuals dying or to measure total biomass lost. The cases of the fern spike and the planktonic foram offer unusual glimpses of death in the geologic past. Thus, we are obliged to return to analyses of broad patterns of victims and survivors—with the accounting done mostly with species, genera, and larger taxonomic groups.

I will not describe the other Big Five extinctions one by one. Lists of winners and losers become dense—even boring—when, as we go back in time, organisms become less and less familiar. I will detail aspects of some of the events, but only when appropriate to more general discussions of the extinction phenomenon. As we shall see, all extinctions are different—in both pace and consequences—but there are patterns that lead to some general answers.

MEASURING EXTINCTION

One way to measure the severity of the K–T extinction is to calculate the percentage of species (or groups of species) living in the latest Cretaceous but dying out at the K–T boundary. The percentages are approximately as shown on facing page: Most striking is the increase in killing as we move down the list, from large groups to small. The small groups are subsets of the larger: each phylum contains one or more classes, each class contains one or more orders, and so on.

To appreciate these numbers, try the following thought experiment. Imagine a world in which each species has ten individual organisms, each genus ten species, each family ten genera, and so on up to a single phylum containing ten classes. The arithmetic works out to precisely one million

Group	Percentage Lost
phyla	0
classes	1
orders	10
families	14
genera	38
species	65–70
individuals	?

individuals. Now suppose that individuals are killed at random, without reference to membership in species or higher groups. This has been called the Field of Bullets scenario—all individuals exist in a field of flying bullets, and death or survival is solely a matter of chance. The image is awful, but it does the job.

If 75 percent of the individuals in this thought experiment are killed, what are the extinction percentages for the taxonomic groups? Starting from the top, the kill rate for phyla must be zero because there is only one phylum and it loses only three-quarters of its members. Each of the 10 classes had 100,000 individuals, so we can ask, What is the probability that any one of these classes will lose *all* 100,000 of its members if the overall death rate is 75 percent? That probability is, for all practical purposes, zero; this makes it surprising that even 1 percent of the classes went extinct at the K–T (1 out of 82 classes).

As we move down the taxonomic hierarchy, extinction of groups by chance alone becomes more and more likely. At the bottom, the probability that all ten individuals in any one of the species will be killed by the Field of Bullets is about one in twenty (.056, or 5.6 percent). Therefore, only about 5 percent of the species will go extinct in our imaginary event. If killing is random, 75 percent of the individuals must thus be killed to cause the extinction of 5 percent of the species.

The main message of this exercise is that, with random killing, the extinction rate will increase down the taxonomic ladder. The extinction rate will be zero at some high level, unless, of course, all organisms are killed. Let me hasten to point out what many readers have probably already noticed: the numbers I used in the thought experiment are unrealistic. As the previous chapter emphasized, sizes of groups in nature are always skewed, small groups being most common.

The Field of Bullets scenario can be used with realistic numbers, and although the mathematics becomes complex, relationships between extinction rates are substantially the same as before. Figure 4-2 illustrates a technique called reverse rarefaction. It is based on the assumption that species rather than individuals are killed at random. The raw data for the analysis came from actual numbers of species, genera, and families in groups of living organisms.

Figure 4-2 confirms the pattern of an increasing extinction rate with decreasing taxonomic level. Actually, the 65–70 percent estimate for K–T species extinction was taken from this graph because the fossil record of species is too incomplete for direct counts. That is, the species estimate is based on the numbers of extinctions at higher levels, using the assumption of random killing of species (Field of Bullets scenario).

Many readers will have heard that 96 percent of all species living near the end of the Permian were killed in the big

Reverse Rarefaction

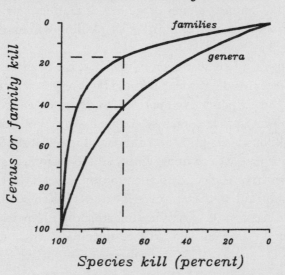

Species kill (percent)

FIGURE 4-2. Method of reverse rarefaction used to estimate percentages of species killed from counts of extinctions at higher taxonomic levels. As an example (dashed lines), an extinction of 40 percent of the genera implies a 70 percent species extinction. The method uses the simplifying assumption that species are killed at random, without regard to membership in a genus. Thus, a pure Field of Bullets scenario is assumed, and the resulting predictions are accurate only to the extent that this assumption is satisfied. The estimate that up to 96 percent of all species were killed at the end of the Permian (Raup, 1979) is based on this graph and the observation that 52 percent of the families died out.

mass extinction at that time. This number comes from the reverse rarefaction graph of Figure 4-2. This estimate is probably an exaggeration, because the extinction of species is not completely random. If extinction is focused on certain genera and families, killing will be concentrated in these groups. If reverse rarefaction is used to estimate species kill from the extinction rates of genera or families, any departure from random (Field of Bullets) killing will exaggerate species kill.

I am slightly embarrassed by the wide use of the figure of 96 percent for the Permian because I was responsible for it in a 1979 article presenting the reverse-rarefaction method. Although my article contained ample caveats about the random-killing assumption and although I said that the 96 percent estimate was an upper limit, all too many users of the number have neglected to mention the caveats. In truth, I probably did not exert myself to emphasize them. The whole question of selectivity in extinction is crucial to the extinction problem and will be discussed further in later chapters.

A Note on Killing

Let me digress briefly to comment on the Field of Bullets scenario and all this talk of killing. The prominence of death and killing in the discussion is slightly morbid. But this follows from my attempt, mentioned in the first chapter, to view extinction as a process more active than has been traditional.

On the other hand, several quieter extinction scenarios have been proposed. They are based on the idea that extinction occurs because the birthrate fails to keep up with the death rate. This is aesthetically pleasing, for no individual is actually killed—extinction is due merely to a lack of sufficient births. Nobody is hurt. Although this scenario is certainly possible, I submit that its main appeal is aesthetic—a common hazard in the development of scientific theories.

This reminds me, in turn, that several paleontologists have recently proposed that the mass extinctions are not that at all—that they are only the passive result of a marked drop in the origination of new species. Reasonable, though by no means compelling, documentation has been offered to support this theory, but I continue to suspect that the driving motivation—unconscious perhaps—is a wish to avoid all the killing. We'll see.

Duration of Mass Extinctions

Did the K–T extinction last several millions of years, or was it over in a few minutes? This is a vexing and important question, to which the fairest answer is that we don't know.

Many published charts of time ranges of fossil lineages leave the impression that K–T extinctions were instantaneous, because lines depicting extinct lineages stop abruptly at the K–T boundary (the dinosaur lineage in Figure 1-1, for example). But these charts are greatly simplified. For most lineages, drawing the line up to the K–T boundary means only that the lineage has been found somewhere in the last recognizable time unit of the Cretaceous—usually the Maestrichtian stage, the last nine million years. The lineages could have died out anywhere in that time. If the extinctions were, in fact, spread over this much time, we cannot call the K–T an instantaneous "event."

In a single outcropping of rock that includes the K–T boundary, paleontologists can collect fossils intensively from each centimeter of rock to see exactly where species disappear as the boundary is approached. If they do this at many sites around the world, they should be able to settle the question of the duration of the mass extinction. Unfortunately, severe practical problems hinder this effort.

Geologic dating is often so uncertain that one cannot be sure that the rocks at different sites are the same age. Even if the K–T boundary can be identified at each site—not always possible—the boundary itself may not be the same age at all sites. Suppose that at one site, the rocks deposited during the last two or three million years of the Cretaceous were eroded away before the start of the Tertiary. The K–T boundary, defined as the upper surface of the youngest Cretaceous rocks, would be several million years older than the actual age of the K–T transition.

This logic is illustrated in Figure 4-3. The rock column on the left represents a complete chronology, with alternating black and white bands to suggest a succession of different kinds of rock. The true K–T boundary, at 65 ma BP, is indicated by the arrow. The column on the right shows what the same sequence might look like if an interval of erosion occurred near 65 ma BP. The wavy line indicates the eroded surface of Cretaceous rocks, on which Tertiary rocks were deposited. Note that a black band seen between 65 and 66 ma BP on the left was removed completely by erosion in the column on the right. When sedimentation resumed, Tertiary rocks were deposited on the erosion surface. As a result, the boundary between the youngest surviving Cretaceous and the oldest surviving Tertiary brings rocks of different ages into contact. In other words, rocks formed well after the end of the Cretaceous are resting on rocks formed well before the end. So, the eroded column

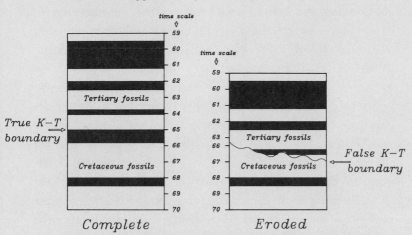

Effects of Erosion

FIGURE 4-3.   Two hypothetical rock sequences showing the effects of erosion on the position of the Cretaceous–Tertiary boundary. In the sequence on the right, several million years of record was lost after the end of Cretaceous time but before the start of Tertiary deposition. As a result, the first Tertiary formation rests on rocks deposited up to two million years before the end of Cretaceous time. The loss of this record causes uncertainty in times of extinction of Cretaceous lineages; a fossil may be present at a false K–T boundary yet not have survived up to the true time boundary.

contains a Cretaceous–Tertiary boundary, but its age, in millions of years, is an artifact of depth of erosion.

Because of erosional and other gaps in the rock record, it is extremely difficult for us to amass a sufficient quantity of high-quality data to say anything more than that the extinctions occurred sometime near the end of the Cretaceous.

A curious feature of several of the mass extinctions is that they appear in parts of the geologic column having unusually long gaps. This is generally true for the K–T extinction and is even more pronounced for the Permian mass extinction. In most parts of the world, large segments of the latest Permian record are completely missing, the gaps often being three million years, or more.

Only in China does the Permian sequence appear to be reasonably complete. (Strangely, China has a significant fraction of the interesting geology of the world.)

The number of big mass extinctions is small, and the association of several of them with long gaps may not indicate anything. But if the association is real, it may be telling us something about causes. It has been suggested that because many of the gaps are caused by drying out of inland seas, the extinctions may have been caused by a global lowering of sea level. I will explore this possibility in a later chapter.

The suddenness of an extinction may be difficult to assess even in a complete sequence. If preservation of fossils is spotty—as it often is—the last occurrence of a kind of fossil may not record the actual time of extinction. The last occurrence really tells us only that the species was still alive at the time of the last preserved fossil.

Let me give an example. A few years ago, I attended a conference on the Biscay coast. We took a field excursion to splendid Cretaceous–Tertiary exposures: at Zumaya, in Spain, and nearby Biarritz, in France. Fossils had been collected intensively for years by a German group and, independently, by an American team. At issue was whether the ammonites—important victims of Cretaceous extinction—died out at the K–T boundary or much earlier. Only one ammonite specimen had been found close to the boundary; all the rest were at least ten meters below. The lone specimen was in rather poor condition, and some people argued that it was an older fossil first deposited well back in Cretaceous time, then eroded (as a pebble) and redeposited. It was thus possible to argue that the ammonites were dead and gone long before the end of the Cretaceous.

The leader of the excursion, one of the German paleontologists, offered a bottle of the best Spanish brandy as a prize to anyone who could find an ammonite within ten meters of the K–T boundary. The boundary, by the way, is clear and sharp in the Biscay cliffs, recognizable by an abrupt change in the color and structure of the rocks. Before the morning was out, one of the group had indeed found an excellent ammonite specimen well within the ten-meter limit. Years of hard collecting by a few dedicated paleontologists had not turned up what many pairs of trained eyes could find in a couple of hours, given sufficient incentive. From this and other finds on the Biscay coast, it is now clear that several ammonite species can be found right up to the K–T boundary. But few situations have had the benefit of such intensive study.

Many paleontologists have strong views on the issue of the duration of mass extinctions. Competent and honest scientists look at the same data and reach opposite conclusions. My own reading, for most of the cases, is that sudden killing best explains the facts. But I may be the victim of unconscious bias. Much depends on which side of the issue is assigned the burden of proof—the observations can be induced to fit reasonably with either side. At any rate, I will continue to refer to the mass extinctions as "events," accepting the implication that they were sudden and short-lived.

Do Mass Extinctions Differ from Background?

Conventionally, the history of life is viewed as a fairly steady flow of change, interrupted occasionally by a livelier pace. Mass extinctions seem to stand out from the normal course of history. But is this distinction real?

An analogy within our experience is the hurricane, certainly something different from the general run of weather. People who live in the West Indies or Japan can recite the recent history of hurricanes in their area. But are hurricanes really distinct from other severe storms? A *hurricane* is defined officially as a tropical cyclone with a pronounced rotary motion and sustained surface winds of at least 64 knots (73 mph). With lesser winds, the same weather disturbance is called a *tropical storm* (34- to 63-knot winds), and this leads down through a series of milder tropical cyclones: the *tropical depression*, with surface winds less than 33 knots, and the *tropical disturbance*, which has no strong winds but still has a reasonably organized cyclonic structure. By this pigeonholing, we create the appearance of discontinuities where none exist.

Figure 4-4 shows a histogram of variation in intensity of extinction over the past 600 million years. To make the graph, the percentage of genera of marine fossil animals going extinct in each of the recognized time intervals was tabulated. Notice first that the histogram is skewed, with low-intensity extinction being most common. The tail extending to the right contains the big mass extinctions. The histogram clearly shows that mass extinctions are merely one end of a fairly smooth distribution.

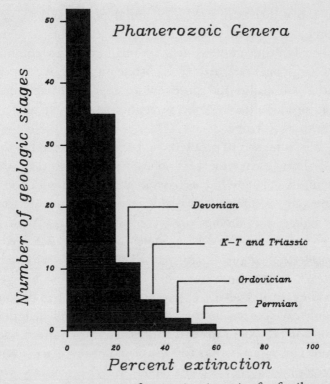

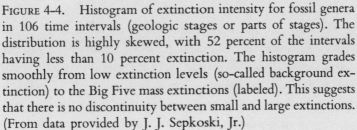

FIGURE 4-4.   Histogram of extinction intensity for fossil genera in 106 time intervals (geologic stages or parts of stages). The distribution is highly skewed, with 52 percent of the intervals having less than 10 percent extinction. The histogram grades smoothly from low extinction levels (so-called background extinction) to the Big Five mass extinctions (labeled). This suggests that there is no discontinuity between small and large extinctions. (From data provided by J. J. Sepkoski, Jr.)

The Big Five seem to stand out as special because extinctions of this magnitude are rare. But there is nothing mysti-

81

cal about their rarity. Many natural phenomena—severe storms, earthquakes, volcanic eruptions, droughts—are distributed in time in the same way: small events are common and large ones rare. At the high-intensity end, we experience few examples and those few seem special. I think this is what produces the common perception of mass extinction as something different.

When this sort of problem has been encountered in other disciplines, by far the best solution is the one that river engineers and hydrologists have adopted to deal with floods. First, they assemble all available records of flow rate for a river. Then they arrange the records so that intensity can be expressed as that flood level which is equaled or exceeded (on average) every so many years. Thus, the 10-year flood is that which occurs (on average) every ten years, the 100-year flood every 100 years, and so on. This time interval, commonly called the *return time* or *waiting time,* is the length of time one can expect to wait for the return of a given flood level. This system works for any phenomenon where events become rarer as they become larger. And this is exactly what we have in extinction.

Hydrologists use a technique called extreme-value statistics to make intelligent guesses of waiting times beyond the length of available historical records—the waiting time for the 1,000-year flood when only 100 years of river levels have been recorded. These estimates are imperfect, but the attempt is important—when even poor predictions are better than nothing.

In extinction, we have a good record for the past 600 million years, so we can define the 10-million and 30-million-year extinction levels with confidence. The K–T extinction turns out to be a 100-million-year event. The extinction at the end of the Permian may be a 600-million-year event, because we have had only one of them in 600 million years. Or it may be the billion-year event. It is also possible that the Permian is only a 200-million event and that we were lucky to have had only one of them in 600 million years. When waiting times approach the length of the record, estimates get dicey.

I once tried extreme-value statistics on extinction data to ask, "How often should we expect extinction of all species on earth?" I don't have much confidence in the results, but they are at least comforting: extinctions sufficient to exterminate all life should have an average spacing of well over two billion years.

The Kill Curve

Figure 4-5, a *kill curve* (my own invention, as you can tell from the name), depicts the average species kill for a series of waiting times. The curve was constructed at the level of species by interpolating data from about 20,000 extinction records of genera.

The events we call mass extinctions are high on the kill curve, whereas what we call background extinction is near the lower-left corner of the graph. There is no indication of any break in the curve that would justify depicting mass extinctions as something different from lesser extinctions. If there is such a discontinuity—and there may be—our fossil record does not show it.

Notice on the kill curve that for waiting times of 100,000

83

years (labeled 10 with an exponent of 5) and less, species extinction is negligible. This means that the typical 100,000-year interval experiences almost no extinction, just as a typical week in a person's life has no severe storms or large earthquakes. The kill curve does not predict when a mass extinction will occur, of course; it only tells us the average likelihood of an occurrence in a given length of time.

The most important message of the kill curve is that species are at low risk of extinction most of the time (the lower part of the curve), and this condition of relative safety is punctuated at rare intervals by a vastly higher risk of extinction. Long periods of boredom interrupted occasionally by panic. Any explanation of the causes of extinction, to be plausible, must accommodate this pattern.

In this chapter, I have hit a few high spots in the complex matter of defining and describing mass extinctions. Mass extinctions in the history of life are surely real even though they are difficult to measure and even though our knowledge has gaps, such as whether mass extinctions last a short or a long time. Sizes of extinction seem to grade imperceptibly downward from mass extinctions through lesser extinctions to what has been called background extinction. Thus, we cannot define mass extinction except by agreeing on an arbitrary cutoff, such as 65 percent species kill, or an arbitrary waiting time, such as 100 million years.

Even if we accept that mass extinctions can be defined only with arbitrary rules, there may still be important discontinuities in the biological consequences of extinction. There may be a species kill above which the fundamental structure of ecosystems breaks down, triggering significant

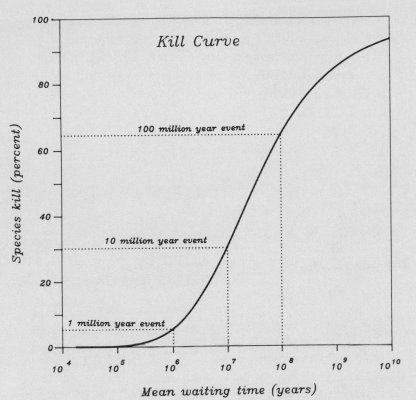

FIGURE 4-5. Kill curve summarizing the history of species extinction for marine organisms over the past 600 million years (Phanerozoic). The curve shows the average spacing (waiting time) for events of varying extinction intensity. For example, a species kill of about 5 percent occurs about every one million years (10 to the 6th power). The Big Five mass extinctions are approximately 100-million-year events (with the possible exception of the Permian extinction) and kill an average of 65 percent of all species. (Based on data provided by J. J. Sepkoski, Jr.)

evolutionary consequences not produced by the accumulation of lesser killing. One could devise interesting and plausible scenarios in this vein, although they would be difficult to document in the present state of knowledge. I will return to this theme in a later chapter.

SOURCES AND FURTHER READING

Alvarez, L. W., W. Alvarez, F. Asaro, and H. V. Michel. 1980. Extraterrestrial cause for the Cretaceous-Tertiary Extinction. *Science* 208:1095–108. The original research article proposing comet or asteroid impact as the cause of the K–T mass extinction.

Gore, Rick. 1989. What caused the earth's great dyings? *National Geographic* 175 (June): 662–99. An elaborate and authoritative treatment of mass extinctions.

Gumbel, E. J. 1957. *Statistics of extremes.* New York: Columbia University Press. A heavy mathematical treatment of the problem of predicting rare events.

Raup, D. M. 1979. Size of the Permo-Triassic bottleneck and its evolutionary implications. *Science* 206:217–18. A research article concluding that up to 96 percent of all species were killed in the Permian mass extinction.

Raup, D. M. 1991. A kill curve for Phanerozoic marine species. *Paleobiology* 17:37–48. A research article deriving the kill curve of Figure 4-5.

Raup, D. M., and D. Jablonski, eds. 1986. *Patterns and processes in the history of life.* Berlin: Springer-Verlag. A collection of research articles from a Dahlem conference held in Berlin in June 1985.

Raup, D. M., and J. J. Sepkoski, Jr. 1984. Periodicity of extinctions in the geologic past. *Proceedings of the National Academy of*

Sciences 81:801–5. The research article claiming that extinction events are evenly spaced in time.

Sepkoski, J. J., Jr. 1986. Phanerozoic overview of mass extinction. In *Patterns and processes in the history of life,* 277–95 (reference given above). The source for Figure 4-1.

CHAPTER 5

SELECTIVITY
OF EXTINCTION

A central issue in the bad-genes/bad-luck question is whether extinctions are selective. Are victims chosen at random from total biodiversity—that is, not chosen on merit—or are certain organisms or certain habitats at higher risk? Are there species with immunity, and, if so, what is the nature of that immunity? The Field of Bullets scenario, discussed in Chapter 4, assumes total randomness. If we knew how close this was to the truth, we would know much more about extinction and its causes. Also, the selectivity issue has an obvious bearing on the role of extinction in evolution. The less random extinction is, the more influence it should have on the course of evolution, for good or ill.

ICE AGE BLITZKRIEG

The Spaniards introduced horses into the New World in the sixteenth century, and legend has it that these large animals made quite an impression on the Indians. But horses were, in fact, not new to North America; they had a long history here but were wiped out several thousand years before Spaniards arrived. From fossil evidence, we know that North and South America had a fully developed fauna of large mammals, including horses, and that this fauna survived through most of the Pleistocene glaciation. Horses had mingled with mammoths, mastodons, scimitar cats (sabertooths), and giant ground sloths. Our zoos today would be exotic indeed if these animals had survived; the closest we can come are museum exhibits of mounted skeletons, especially at Page Museum, located on the edge of the La Brea tar pits, in Los Angeles.

Pleistocene mammal extinction is a good example of selectivity. Because carbon-14 dating is good, we have a far more accurate chronology for this than for older extinctions. The extinctions show interesting patterns in time and space. For example, the continents of the Americas, along with Australia and Madagascar, were much harder hit than other areas. Nor were the extinctions synchronous. Large marsupials disappeared from Australia several thousand years before their counterparts in North and South America; the extinctions of giant lemurs in Madagascar occurred at yet another time. The woolly mammoth died out about 18,000 yr BP (years before present) in China, 14,000 yr BP in Britain, 13,000 yr BP in Sweden, and sometime after

12,000 yr BP in Siberia. Many of the extinctions were total (that of sabertooth cats, for example), whereas others (those of horses and camels) occurred on only one or two continents, leaving survivors elsewhere.

In North America, carbon dating of Pleistocene extinctions converges on a narrow range from 10,800 to 11,000 yr BP (8,800 to 9,000 B.C.). This is shortly after the earliest good evidence of human habitation in North America (Clovis culture), and some archaeological sites show that early man hunted and butchered the large mammoths. These sites have given rise to the so-called blitzkrieg theory of overkill by human hunters as an explanation for the extinctions. In the blitzkrieg model, geographic differences and lack of synchroneity are interpreted as caused by patterns of human habitation. Areas of long, continuous habitation, such as Asia and Africa, show less extinction than areas occupied by humans late and rather suddenly, like North America.

The blitzkrieg theory is controversial, and much has been written on both sides. Its most articulate spokesman—and careful analyst—is Paul Martin of the University of Arizona. Martin has written extensively on the problem and is always careful to present opposing views.

Although many people (not including Paul Martin) have called the Pleistocene wipeout of mammals a mass extinction, it was not nearly as pronounced as that. The dramatic losses were limited to mammals, some of the large flightless birds, and a few other groups. Nothing unusual happened to marine animals. Only the most chauvinistic members of the Mammalia—meaning us—could see that the event as remotely similar to the K–T mass extinction or others of the Big Five. Nevertheless, the sudden removal of most large herbivores and carnivores from vast continental areas must have had substantial effects on some terrestrial ecosystems.

SELECTIVITY OF THE BLITZKRIEG

The Ice Age extinction was selective in that it affected mammals far more than other organisms and in that extinction rates were far higher among large mammals than small. The cutoff for distinguishing large and small mammals is conventionally set at 100 pounds (44 kg) adult body weight.

Selection for size can be seen among genera as well as species. The following figures are for North American mammals:

	Number Living before the Extinction	Number Dying	Percentage Dying
Small animals			
Species	211	21	10%
Genera	83	4	5%
Large animals			
Species	79	57	72%
Genera	51	33	65%

The higher percentages for large animals are indeed striking, and sample sizes are large enough to make statistical testing possible. Testing shows that the preponderance of extinction among large mammals is not likely to be due merely to chance (bad luck). It appears that large size really did put land mammals at a much higher risk of extinction.

91

As was already noted, the cause is thought by Paul Martin and others to have been overhunting by prehistoric people.

If Martin's theory is true, Pleistocene extinction, with its human intervention, may not help much with the more general problem of extinction. The late Robert MacArthur, a brilliant geographic ecologist at Princeton, once wrote that man is the only species with sufficient intelligence and concentration span to cause, by predation alone, the total extinction of another species if the prey species is geographically widespread.

Many paleontologists have taken issue with the blitzkrieg theory. Criticism generally takes two forms. First, geologic dates critical to the case have been challenged. Radiocarbon dating, like all dating based on radioactive isotopes, is subject to error and varying interpretations. Thus, legitimate debate often attends controversial conclusions based on radiocarbon dating. Much of Martin's argument depends on dates of early man in North America.

The second attack on blitzkrieg argues that climatic instability is a more likely cause. A general melting of ice sheets caused a rising sea level and a drying of glacial lakes. The resulting ecological disruption could have been too much for the large mammals. In fact, some have used this argument to explain the apparent coincidence in timing between the extinctions and arrival of people in North America by noting that human immigration may have been made possible by an amelioration of climate.

The climatic argument has always seemed to me to be somewhat ad hoc and certainly not as neat and simple as the overkill theory. But blitzkrieg has also been criticized on grounds that it is too simple. In my experience, about as many people say, "Scientific problems rarely have simple answers," as say, "Where there is a choice, simple explanations are most likely to be correct." Both statements are rhetorical rather than analytical, and one hates to see them used as arguments for or against a theory.

BODY SIZE AND THE K–T EXTINCTION

In preparing this chapter, I dug into recent literature to find out whether body size was a critical factor in the Cretaceous mass extinction. It is commonly accepted that large animals were discriminated against, but I wanted to get some details. I found the following statement in a general review of body size in evolution, published in 1986 by Michael LaBarbera (University of Chicago):

If large-bodied organisms tend to be stenotypic [restricted to a narrow range of environments], then one might expect differential survival of large- and small-bodied groups in times of mass extinction. This certainly seems to be the case for the terminal Cretaceous extinction event where *virtually all large-bodied vertebrates were eliminated.* (italics added)

Next, in a review of K–T extinctions also published in 1986, by William Clemens, a Berkeley paleontologist specializing in terrestrial vertebrates, I found this statement:

Whether average adult or individual body size is considered, a particular range of *body size is not a common feature of all groups that survived or became extinct at the end of the Cretaceous.* (italics added)

93

These statements illustrate some of the problems of diagnosing selectivity in extinction. Both statements are true. In support of Mike LaBarbera, the largest vertebrate animals living in the late Cretaceous did go extinct—meaning large reptiles—and this suggested to him, quite validly, that body size may have been a factor. Bill Clemens, on the other hand, was using a size cutoff of 25 kilograms for the distinction between large and small—analogous to the 44-kg cutoff used in the Pleistocene example. On this basis, Clemens found that all taxonomic groups that survived the K–T event contained at least some species with adult body weights above the established cutoff. These included a number of very large crocodiles and turtles that survived without apparent difficulty.

Is either view definitive? All is in the eye of the beholder. Good arguments can be made for either approach to assessing body size. Should average or maximum size be used? For that matter, is body weight the best measure of size? Body length or height is often used instead. Also, Clemens has emphasized that it is important to decide whether size at birth is more or less significant than adult size. Even after one decides these questions and proceeds to amass a representative set of measurements, the choice of technique for analyzing the data contains many judgment calls.

Because of the large number of choices available in planning a research strategy, a case must be extremely clear-cut in order for all strategies to come up with the same answer. Body size at the K–T is evidently a sufficiently close judgment that different approaches can and do produce radically different answers.

OTHER EXAMPLES OF SIZE BIAS

There are several other cases of alleged size bias in extinction. The ammonites, as a group, contained some of the largest invertebrates of all time. Some were several feet in diameter. All ammonites died out at the end of the Cretaceous, but the largest ammonites were no longer around by then. A similar example is found with extinct Paleozoic arthropods called eurypterids. These strange creatures, somewhat like giant crabs, were also among the largest invertebrates ever to have lived, but there is no evidence that large size caused their extinction. Again, the largest specimens are not found at the end of their geologic range.

If we were fortunate enough to prove beyond reasonable doubt that large body size does indeed correlate with extinction, explanations from theory would be readily available. As Mike LaBarbera and others have shown, body size has serious physiological consequences. It affects aspects of the strength of muscles, bone, and tendons, as well as metabolic rate (which generally goes down as body size increases). Numerous ecological and demographic variables are also closely tied to body size. Large animals have smaller populations and are more dispersed over their geographic ranges. Moreover, they tend to put less of their total energy into reproduction. In view of all this, it is a little surprising that the extinction-size correlation is not stronger.

OTHER EXAMPLES OF SELECTIVITY

Size is not the only trait that suggests a proneness to extinction. It is commonly held, for example, that tropical organisms are more likely to go extinct than their relatives in cooler climates. Planktonic organisms are said to be at greater risk than bottom-dwelling aquatics, and marine reef communities more vulnerable than nonreef communities.

My own feeling is that most of these claims are not worth a damn! Sadly, to test such claims is nearly impossible. Let me explain. Suppose we are studying one particular extinction event and have a list of victims and survivors. Such lists tend to be rather short, especially if we are working at a high taxonomic level (order, class, family). Furthermore, the best studies are done by experts specializing in one group of organisms, thus limiting the lists of victims and survivors even more. Small numbers make statistical testing tricky.

Once we have the lists, we must search for common denominators: characteristics shared by most victims but not survivors, or vice versa. This is straightforward, and we have seen the results in the case of mammalian body size. *The problem is that organisms have a virtually unlimited number of characteristics that might be important:* anatomical, behavioral, physiological, geographic, ecological, and even genealogical. We can compare lists of victims and survivors with as many different traits as we have energy. If the lists are not too long, it becomes virtually inevitable that we will find one or more traits that match the lists closely enough for us to make a case.

If we find an interesting correlation by this procedure, we

can apply standard statistical tests to evaluate the possibility that the correlation is due to chance alone. Each such test asks, in one way or another, "What is the probability that the random sprinkling of a particular trait among species would, by chance, yield a correlation as good as the one we observe?" If that probability turns out to be very low—say, 5 percent or less—we feel comfortable in rejecting random sprinkling and concluding that the observed correlation is true cause and effect.

The fatal flaw in this logic is that testing cannot be adjusted for the fact that we tried many traits before finding a promising one. Remember that one out of every twenty completely random sprinklings will, on average, pass our test if odds of twenty to one are considered acceptable— as is common in scientific research. Because it is virtually impossible to keep track of the number of traits we have considered—many were discarded at a glance—we cannot evaluate the test results for any one trait.

This problem is not unique to paleontology, or to science either. If you have difficulty accepting my reasoning, try some experiments yourself. Take some baseball statistics or election results or anything that will provide lists of winners and losers. Fifty or a hundred results should be adequate. Then inspect the lists to see what characteristics the winners or the losers have in common. The pattern does not have to be perfectly consistent—a statistical tendency is enough— and you are free to change the ground rules as you go along. You can even redefine winner and loser if this will help. Pay special attention to the smaller category of outcomes. For example, you may wish to compare characteristics of first-place baseball teams with those of all other teams. The shorter list (first-place teams) is more likely to have things in common than the longer list. If so, you may be able venture conclusions like "Most managers (or all, if you are lucky) of first-place teams are firstborns, whereas managers of other teams follow the national average." Here is my own tongue-in-cheek attempt, not having to do with baseball:

If cities of the world are measured by population, some are "winners" and some "losers." Research has shown that people are strongly attracted to cities with names starting in the second half of the alphabet. Although this relationship is quite obvious in any good demographic data base, it can be proven by a simple statistical test, as follows. According to the World Wide Atlas *(Reader's Digest, 1984), the seven most populous metropolitan areas are (in descending order):*

> *Tokyo-Yokohama*
> *New York*
> *Mexico City*
> *Osaka-Kobe-Kyoto*
> *São Paulo*
> *Seoul*
> *Moscow*

Note that all of these names start with letters in the M–Z range, virtually the second half of the alphabet. Even the smaller cities contributing to the seven metropolitan areas are in this range (Yokohama) or close to it (Kobe, Kyoto).

It could be argued that this near-perfect correspondence is an accident of random sprinkling of populations and city names. It is also possible that most cities in the world have initial letters late in the alphabet. To test these alternative hypotheses, a control sample was taken from the same source. The next seven most populous metropolitan areas are (in order):

> *Calcutta*
> *Buenos Aires*
> *London*
> *Bombay*
> *Los Angeles*
> *Cairo*
> *Rio de Janeiro*

With the one exception of Rio de Janeiro, all have initial letters drawn from the A–L part of the alphabet, and even the second and third words of that one exception have initial letters from the A–L range (de Janeiro). Note also that the second words of two other names in the control list have initial letters from very early in the alphabet (Aires and Angeles). The statistical likelihood that this was caused by chance alone is so small that rejection of a hypothesis of randomness is routine. Cause and effect is clearly indicated.

Although further research is needed, it appears most likely that names late in the alphabet give an impression of wealth and plenty, thereby attracting large immigrant populations. Although Rio de Janeiro is an exception, there is a strong possibility that this city has been misnamed or its population underestimated. With regard to this case, it is also significant that the alphabet-population relationship is pronounced only at the high end of the population scale—that is, metropolitan areas with populations approaching or exceeding ten million. Below this level, city names tend to be mixed, probably indicating that these cities have not yet reached a stable demographic equilibrium. Rio de Janeiro, being at the bottom of the list of fourteen top population centers, is near the detectable limit of the alphabet-population phenomenon.

EXTINCTION: BAD GENES OR BAD LUCK?

See how easy it is to establish a case for selectivity? Should this make us skeptical about the claim that large mammals in the late Pleistocene suffered higher-than-average extinction because they were large? How many other characteristics of mammals were investigated? Perhaps the real culprit was diet or thickness of tooth enamel. My own hunch is that body size did matter for Pleistocene mammals, but I cannot prove it.

Taxonomic Selectivity

About one kind of selectivity I have no doubt—some taxonomic groups have significantly higher extinction rates at certain times than can be explained by chance. Consider the dinosaurs. Bill Clemens, the Berkeley paleontologist mentioned earlier, has accumulated large quantities of excellent data on extinctions of vertebrate animals found in the latest Cretaceous rocks in the western interior of North America. In one review, he gives numbers for genera, shown on the facing page.

The overall average extinction rate, 43 percent, is only slightly above the global average for genera at the end of the Cretaceous. For several groups listed above, the samples are small and irregularities may well be due to chance. For example, of nine placental mammals, one went extinct (11 percent) and of four marsupials, three went extinct (75 percent). The big difference in percentage gives the impression that marsupials were much harder hit than placentals (our own ancestors), and of course they were. But the numbers are much too small for statistical confidence, because we cannot distinguish these results from what we might get by

	Number Living before the Extinction	Number Dying	Percentage Dying
"Fish"			
Chondrichthyans	5	3	60%
Osteichthyans	13	5	38%
Amphibia	12	4	33%
Reptilia			
Chelonia	18	2	11%
Eosuchia	1	0	0%
Crocodilia	4	1 ?	25% ?
Eolatacertilia	1	1	100%
Lacertilia	15	4 ?	27% ?
Serpentes	2	0 ?	0% ?
Saurischia	8	8	100%
Ornithischia	14	14	100%
Pterosauria	?	?	100%
Aves	?	?	?
Mammalia			
Microtuberculata	11	4	36%
Marsupialia	4	3	75%
Placentalia	9	1	11%
TOTALS:	117	50	43%

rolling dice, for example, with the same overall odds.

Some disparities, however, cannot be explained by chance. The two dinosaur groups—Saurischia (lizard-hipped) and Ornithischia (bird-hipped)—had a combined total of twenty-two genera, and all twenty-two died out. With a 43 percent overall extinction rate, the probability that 100 percent of any group of twenty-two went extinct by chance alone is virtually nil. This is true even after adjusting for the fact that we chose to test the case of the dinosaurs after we knew that they all had died out. This means that the dinosaurs did something wrong. That is, they had some extra susceptibility to whatever caused the Cretaceous extinctions.

This sort of situation is common, though by no means overwhelming, in extinction. In massive statistical analyses on large data bases for fossil marine animals, I have found a few more cases of apparent taxonomic selectivity than would be expected to have occurred by chance alone. The surplus is not large, but it is real.

THE TRILOBITES' BAD GENES

In the first chapter, I commented that the title of this book was taken from a research article I had published on the extinction of trilobites. The case provides another example of taxonomic selectivity.

In rocks of the Cambrian period (570–510 ma BP), somewhat more than six thousand species of trilobites have been found and named. This is three-quarters of the fossil species known from the Cambrian. By the end of the Paleozoic era, 325 million years later, all were gone. On the working assumption that speciation and extinction rates for trilobites were the same as for all other animals of the Paleozoic, my question was whether a group as large as the trilobites could have drifted to extinction by bad luck—as an affluent gambler can drift to bankruptcy, given enough time.

I used mathematical models (designed along the lines of the random walks described in Chapter 3) to estimate the probability that the trilobites could have died out because of a chance excess of species extinctions over speciations. The result was a vanishingly small likelihood that chance was operating alone in the trilobite case. The working assumption that trilobites had inherent extinction and speciation rates equal to the Paleozoic averages was clearly wrong. It followed that trilobites had (for some reason) either less capability for speciation or a higher risk of extinction. Testing the latter possibility, one finds the extinction credible only if one assumes life spans of trilobite species 14–28 percent less than the Paleozoic average.

From this, I concluded that the trilobites were indeed doing something wrong (or that the other groups were doing something better). One vote for bad genes. This analysis does not tell us, of course, what the trilobites did wrong or what the other animals did better. But it is a start. It assures us of a real discrepancy to be explored.

SOME IMPLICATIONS

If some biological groups are really harder hit than others, stresses causing the extinction must be "seeing" traits shared by the species in those groups. Because members of the groups are, by definition, related to each other by common ancestry, they inevitably share some characteristics. For example, all members may have about the same metabolic rate, or size, habitat preference, or geographic range, to mention but a few of the possibilities.

This reasoning moves us a step closer to an understanding

of why dinosaurs died out at the end of the Cretaceous and mammals did not, even though the two had nearly the same number of species and genera. Dinosaurs evidently shared one or more characteristics that made them more vulnerable than mammals.

This does not mean that dinosaurs were necessarily inferior or ill adapted. After all, they had been successful for 100–150 million years. No Mesozoic biologist could have predicted their demise.

Tom Schopf, my colleague at Chicago until his untimely death a few years ago, proposed an interesting cause for dinosaur extinction. He surveyed all late Cretaceous dinosaur occurrences and noted that, in many parts of the world, it was not possible actually to prove that the dinosaurs had still been living at the very end of the Cretaceous. Only in the western interior of the United States and Canada was there clear evidence for living dinosaurs just before the K–T boundary. Tom theorized that some fairly localized event or environmental perturbation in western North America killed dinosaurs while mammals were protected by having species living in other parts of the world.

My vertebrate paleontologist acquaintances have serious doubts that the dinosaurs were actually restricted to North America near the end of the Cretaceous. Being outside the field of vertebrate paleontology, I have no good basis for judging. But if Tom's idea should be correct, we would be faced with a strange mixture of bad genes and bad luck. Dinosaurs had the bad luck to occupy a region where they were vulnerable. But perhaps their genes were also at fault, as a failing of their inherent ability to migrate and maintain a global distribution. The reasoning quickly becomes murky, and I don't want to claim that we are even on the right track. Nevertheless, I think it is clear that something as simple as geographic distribution could easily mean the difference between survival and extinction.

Another Chicago colleague, Dave Jablonski, did a statistical analysis of fossil mollusks living during the final sixteen million years of the Cretaceous. He found that before the big mass extinction, species and genera with large geographic ranges resisted extinction better than did those with small ranges. But, for species at least, this relationship breaks down at the K–T mass extinction. That is, wide geographic distribution offered protection for species during quiet times but not during the mass extinction. Perhaps the stresses that caused the mass extinction were so widespread that there was no safety anywhere. For genera, however, Jablonski found that wide geographic range still provided some protection at the mass extinction. Thus, selectivity of extinction can vary with level in the taxonomic hierarchy.

SUMMARY

Extinction is apparently selective to some degree and in some instances, but proving this is devilishly difficult. Whatever the actual level of selectivity, it is never very prominent. At every turn in the search for good cases, we are blocked by ignorance. Or we run the risk of letting our enthusiasm for answers cloud good sense. But this adds to the fun and challenge of the whole enterprise. How nice it is to be in one of the difficult sciences—as opposed to one of those commonly called the hard sciences!

SOURCES AND FURTHER READING

Clemens, W. A. 1986. Evolution of the vertebrate fauna during the Cretaceous–Tertiary transition. In *Dynamics of extinction,* ed. D. K. Elliott, 63–85. New York: Wiley-Interscience. A source for the discussion of taxonomic selectivity of the K–T extinction.

LaBarbera, M. 1986 The evolution and ecology of body size. In *Patterns and processes in the history of life,* ed. D. M. Raup and and D. Jablonski, 69–98. Berlin: Springer-Verlag. A review of the general problem of body size in evolution.

MacArthur, R. H. 1972. *Geographical ecology.* New York: Harper & Row. A monograph on the general subject of the geographic distribution of species.

Martin, P. S. 1986. Refuting late-Pleistocene extinction models. In *Dynamics of extinction,* ed. D. K. Elliott, 107–30. New York: Wiley-Interscience.

Martin, P. S., and R. G. Klein, eds. 1984. *Quaternary extinctions: A prehistoric revolution.* Tucson: University of Arizona Press. Thirty-eight research articles and reviews on all aspects of the Pleistocene extinctions, but emphasizing mammals.

Schopf, T. J. M. 1982. Extinction of the dinosaurs: A 1982 understanding. In *Geological implications of impacts of large asteroids and comets on earth,* ed. L. T. Silver and P. H. Schultz. Geological Society of America, Special Paper 190. Pp. 415–22. Boulder: GSA. A research article arguing that dinosaurs lived only in North America during the latest Cretaceous, thus eliminating the need for a global cause of their extinction.

CHAPTER 6

THE SEARCH
FOR CAUSES

Paleontologists in the past paid surprisingly little attention to what lies behind the many species extinctions documented in the fossil record, but that changed with the Alvarez article in 1980 on planetary collision. Now fierce arguments about mechanisms of extinction pervade the research community. In this short chapter, I will discuss the search for causes in general terms, leaving for later an exploration of details.

THE RARITY OF EXTINCTION

Geologists and paleontologists usually follow the dictum "The present is the key to the past." Many a great advance has been made by a study of present-day analogues because processes can actually be seen in action. But the present helps little with extinction because the disappearance of a well-established species through natural causes is a rare event on human time scales.

The average life span of species in the fossil record is about four million years. Thus, only about one species in four million dies a natural death each year. If forty million species live today, only ten will go extinct in an average year. Species extinction without human influence is rare. At a time of heightened public awareness of endangered species, this may sound surprising, even callous. Even so, the chances of a field biologist's catching a species at the instant of global extinction are small.

It is also important to note that mass mortality over part of a species's range is much more common than total extinction. For example, in the early 1980s in the Caribbean, the common black sea urchin, *Diadema,* was devastated. Mortality rates in many places exceeded 95 percent. The cause was probably a fast-moving, water-borne virus. In a few years, the sea urchin recovered, comfortably avoiding extinction. Cases like this inevitably leave the impression that natural species extinction is more common than it really is.

The estimate of ten species extinctions per year is based on life spans in the fossil record and thus applies to those species sufficiently abundant to be preserved and discovered. This generally means species that achieved a fairly wide geographic distribution, large populations, and a reasonably long duration. Missing are the many species that never got fully established. If these species could be added to the calculations, we would no doubt judge extinction more common—but still rare from the viewpoint of the practicing field biologist.

A biologist studying extinction can take other paths. One is to study extinction locally and then extrapolate to whole species. Within a small area, perhaps a pond or an acre of woodland, species continually appear and disappear. A pond may be full of frogs one year and have none the next. By finding out how local extinctions occur, we may generalize to whole species extending over continents. However, extrapolation from a small part to the whole has serious pitfalls. Conditions that eliminate local populations—a late spring freeze, for instance—may be ineffective, or highly unlikely, over the entire range of the species.

Another approach—gimmick, if you like—is to monitor population size or geographic extent of a species and look for trends suggesting impending extinction. If a species has been declining for several years, one can postulate that extinction is inevitable. The reasons for the decline may tell us something about the causes of extinction. Obviously, this approach is risky because trends often reverse. The Caribbean sea urchin's dramatic decline did not continue and did not lead to extinction. (Yesterday, the Dow Jones industrial average dropped thirty-five points. If this trend continues, at thirty-five points a day, the New York Stock Exchange will be extinct in less than three months.)

Yet another approach is the study of human influences. Examples of man-made extinctions are available, and much can be learned from them. But again, caution is required because the strategy is based on a tacit assumption that environmental stresses produced by human activities also occur in nature. Some of these stresses do occur but others do not.

Recall Robert MacArthur's conclusion that man is the only species able to eliminate an established species by predation.

Given the rarity of natural extinction, we have seen three ways of using proxy data: (1) data on extinction at the local level, (2) data on trends of decline, and (3) data on human effects. All are valid substitutes as long as their proxy nature and the accompanying risks are well understood.

In studying the history of life, we must also keep in mind that some extinction mechanisms have almost surely *not* occurred in human experience. Our time here has been short—a few thousand years of recorded history—and life has spanned 3,500 million years. We have experienced, so far, only 0.0001 percent of the history of life. In fact, it is rather arrogant to think that our 0.0001 percent of life's history should yield a complete sample of nature's processes. Perhaps the past is actually the key to the present (and future).

JUST SO STORIES

Part of the fun of science lies in thinking up theories—explanations for things. Many theories are little more than hunches flowing from "What if?" questions. Some can be discarded almost out of hand as totally implausible or as contradicting some law of nature.

But if a hunch passes initial screening, we may elevate it, privately at least, to the status of a hypothesis. It then must pass more formal tests before it is acceptable as theory, let alone fact. In some cases, the testing procedures are standardized—like some statistical tests—because they have been used many times in similar situations. Or there may be sequences of logical argument that are known to give good results. Surprisingly often, though, the hypothesis to be tested is in a new form, and we must devise new testing procedures.

If a proposed explanation for something passes all tests of plausibility and credibility, is it the correct explanation? Or, if several alternative explanations have been proposed and one is found to be more convincing than any of the others, is that one necessarily correct? My answer to both these questions is an emphatic no.

Something is not proven correct merely because it is shown to be plausible. And this is why circumstantial evidence carries little weight in a criminal court. Merely because a defendant could have committed the crime, by virtue of having the opportunity and perhaps even a motive, we cannot say that the defendant is guilty. Many explanations of extinction are based solely on arguments of plausibility, proposing ways that extinction could have occurred. These are often granted the cynical label of Just So Stories, in honor of Rudyard Kipling's yarns about the origin of the elephant's trunk and the tiger's stripes.

What if the proposed explanation is the best among many? Suppose we have four competing explanations, labeled A, B, C, and D. Suppose further that we have the power to assess accurately the likelihood that each is correct. Let the chance that A is correct be 40 percent and the chances for the other three be 20 percent each. Explanation A is twice as likely as any one of the others. This is fine and may give us some hope for A. But note that the odds are sixty to forty *against* A's being correct. Thus, we cannot select one hypothesis merely because it is better than any of the alternatives. However, if one is better than all the others combined, we have a case. It is a question of plurality versus majority.

The scientific literature, including that dealing with extinction, contains a surprising number of popular arguments based solely on the "better than any of the others" logic.

BEWARE OF ANTHROPOMORPHISM!

Various authors, including news reporters, have made lists of causes that have been suggested for the big mass extinctions. The lists have been intended to show the enormous range of possibilities as well as the silliness of some of them. They do both. "The dinosaurs saw a comet coming and died of fright." But when the silly ones and the Just So Stories are eliminated, a core of stalwarts remains. I think most of my colleagues would agree on the following as serious candidates, not listed in any particular order:

- climatic change, especially cooling and drying
- sea level rise or fall
- predation
- epidemic disease (a kind of predation)
- competition with other species

I have purposely left out, though only for the moment, the environmental effects of comet or asteroid impact.

Each item just listed is reasonable. But I see more than a hint of anthropomorphism here. What are the traditional worries and concerns of people in their own daily lives? The weather, especially cold and lack of rain; water levels (flooding or drying up of rivers and lakes); attacks by wild animals (including insects) or by other people or nations; infectious diseases; and competition (with each other or with other nations). Could it be that the list of probable causes of extinction is simply a list of things that threaten us as individuals? Recall the Four Horsemen of the Apocalypse, representing conquest, war, famine, and plague.

For me to sustain a claim of anthropomorphism, I must be able to draw up a parallel list of plausible but unpopular candidates that do not reflect normal human fears, or at least did not until very recently. Try these:

- chemical poisoning of ocean waters
- changes in atmospheric chemistry
- rocks falling out of the sky
- cosmic radiation
- global volcanic activity
- invasion from outer space

All have been proposed as agents of extinction at one time or another but not taken seriously by the scientific community. The first two are very much on our minds these days, but concern about them is of recent origin and, I submit, has not yet contributed significantly to the canonical list of extinction mechanisms.

The rocks falling out of the sky refers to the idea of a comet/asteroid impact. One reason this notion has caused such a storm of controversy is that it is beyond our experience. Most of us were taught in school that, except for flukes like Meteor Crater, in Arizona, large rocks simply do not fall from the sky. The last three candidates on my list— cosmic radiation, global volcanism, and invasion from space—seem farfetched, and some scientists would try to argue them away on theoretical grounds. But perhaps they merely seem strange (and unlikely) because, like meteorite impacts, they lie outside our experience.

I will let the reader decide whether I have a case for anthropomorphism. At the very least, I think we tend to propose extinction mechanisms taken from among the physical and biological factors that are most familiar. Although to some degree inevitable, it constitutes a bias—a conflict of interest—to be be watched.

THE KILL CURVE REVISITED

In Chapter 4, I introduced the kill curve, deduced from extinction records of fossils. I reproduce it here for convenience as Figure 6-1. The kill curve tells us how many years, on average, we must wait before detecting an extinction of a given magnitude. For example, an average of a million years must pass before an extinction of 5 percent of all living species occurs. This is another way of saying that 5 percent extinctions occur with an average spacing of a million years. As we discussed earlier, larger extinctions are rarer still, so the waiting time for a 65 percent extinction (comparable to the K–T event) is 100 million years.

An important aspect of the kill curve is that it describes killing that occurs in intervals short compared with the enormity of geologic time. For technical reasons, a standard interval of 10,000 years was used to calculate the curve, but this does not imply that the extinctions actually took that long. As we noted earlier, we really don't know how long they lasted.

The kill curve provides an interesting perspective on the

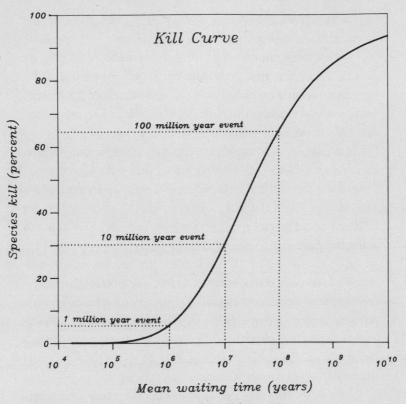

FIGURE 6-1.   Kill curve reproduced from Chapter 4.

killing process and serves to limit our choices of potential causes. The following logic illustrates one way this works. If the risk of extinction for all species was constant throughout geologic time, we should expect some variation in actual extinction, merely as a matter of chance. The average extinction rate for the Phanerozoic is 0.25 percent per 10,000 years. In a purely random model, some 10,000-year intervals should experience more dying, others less. But we could never, by chance variation with a fixed risk, expect even as many as 5 percent of all species to die at the same time. Therefore, the fact that life has experienced 65 percent kill in a short interval is significant. It tells us emphatically that the risk of extinction is not fixed. Species are reacting to some sort of common problem—something that has raised the risk of extinction over a broad front.

This enables us to say that some extinction scenarios cannot be applied to big, sudden extinctions. As an example, consider a virus that is lethal to its host but attacks only that one species. The HIV virus causing AIDS in humans comes close to being species specific. Such viruses are common and have the potential to annihilate a species—several close calls have been recorded in animals. But this extinction mechanism affects only one species at a time. We could not possibly explain a mass extinction by means of species-specific viruses unless we were to postulate either a sudden evolutionary radiation of many different viruses, each attacking a different species, or the appearance of a single virus that attacked large numbers of unrelated species.

Here is another kind of species-specific cause of extinction. The island of Bermuda is home to a large hermit crab; the species is typical in that it requires an empty snail shell to live in. But there are no living snails in Bermuda today large enough to accommodate the crab. Therefore, it uses the shells of a large, extinct snail whose fossils erode out of Pleistocene deposits. The fossil shells constitute a nonrenewable resource for the hermit crab. When the fossils are gone, the crab will either adapt (becoming smaller, I suppose) or die out. If it dies out, the extinction will have been due to a species-specific stress.

Although species-specific causes may serve to explain ex-

THE SEARCH FOR CAUSES

tinctions on the lowest part of the kill curve, they will not explain the large events. For the big events, some sort of common cause is essential. As we will see in subsequent chapters, the common cause may be biological, such as the collapse of large ecosystems, or physical, such as a marked deterioration of climate or a meteorite impact.

SOURCES AND FURTHER READING

Chaloner, W. G., and A. Hallam. 1988. *Evolution and extinction.* London: Royal Society. A collection of research articles on extinction, based on a meeting in London celebrating the two hundredth anniversary of the Linnean Society.

Ehrlich, P. R., and A. H. Ehrlich. 1981. *Extinction: The causes and consequences of the disappearance of species.* New York: Random House. A popular treatment of the extinction problem, with emphasis on contemporary extinction.

Nitecki, M. H., ed. 1984. *Extinctions.* Chicago: University of Chicago Press. A collection of review articles on extinction— past, present, and future.

Stanley, S. M. 1987. *Extinction.* New York: Scientific American Books. A comprehensive treatment written for a general audience; emphasis on climate as a major factor in extinction.

CHAPTER 7

BIOLOGICAL CAUSES
OF EXTINCTION

Almost any stress, physical or biological, can cause extinction. In this chapter and the next, I will review those causes that might be major elements in the extinction process. The discussion will not be limited to mass extinctions, because these events, though important, account for only a small fraction of all extinctions. Calculations from the kill curve show that the Big Five contain fewer than 5 percent of the species extinctions in the Phanerozoic.

I will try to separate physical and biological causes of extinction, but it must be borne in mind that the death of species is ultimately a biological matter. Individual organisms and whole species cease to function as living entities, whether their problems are biological in origin or purely physical. If an animal—say, a squirrel—dies because a boulder falls on it, we might say the death had a physical cause, having nothing to do with biology, but we could just as well claim that the cause was biological, because the squirrel did not have the wit to see the boulder coming. Despite the ambiguities, some explanations of extinction, like virus epidemics, favor biological factors, and others, like climate change, favor physical factors.

ARE SPECIES AND ECOSYSTEMS FRAGILE?

Almost everyone in our culture is taught that plant and animal communities are delicate—intricately balanced yet vulnerable networks of dependencies and interactions. Each species has a place and a role in the community, earned through millions of years of adaptation and coordinated with the evolution of other species. The community is even thought by some to evolve and adapt as if it were an organism itself. Remove one piece of this complex network and other pieces will be lost, perhaps the whole thing. Sudden disturbance, whether by man or by nature, is a negative force, a force of destruction to be avoided.

This view of nature, deeply ingrained and continually reinforced in classrooms, nature study programs, and TV documentaries, is certainly exaggerated. Although thousands of interdependencies are well known—and important—plant and animal communities are less orderly and more resilient than the stereotypical view suggests. Many communities even require some level of disturbance to prosper. A well-known example is the jack pine *(Pinus banksiana)*, which sheds seeds for germination only after subjected to the high temperatures of a forest fire.

Debate has raged among professional ecologists for several years over the structure of plant and animal communities. At one extreme are those who come close to the community-as-organism idea, stressing theories calling for considerable organization among the species living in one place. At the other extreme are those ecologists who see communities simply as collections of species whose habitats happen to coincide or overlap, with each species coping as best it can—eating whatever is available and finding living space opportunistically.

This debate has been constructive and has stimulated thinking about the conceptual foundations of ecology. At times, however, it has been rancorous. My own problem is that the principal spokesmen and intellectual leaders of the opposing sides are two of my favorite scientists: Jared Diamond of UCLA and Dan Simberloff of Florida State University. Although neither takes the most extreme position, Diamond champions the highly organized community and Simberloff spearheads the reaction against it. Both are among the brightest and best researchers the scientific establishment has to offer.

The question of whether species are fragile or resilient is relevant to the problem of extinction in the geologic past. If species are fragile—always vulnerable to extinction—the stresses that cause extinction can be relatively mild and even ordinary. But if species are resilient, conditions must be more severe (and probably unusual) to cause extinction. If natural communities are delicate networks of interdependencies, the loss of one species may cause the loss of others. But if communities are not highly integrated, species extinctions may be independent of one another.

The Case of the Heath Hen

The extermination of the heath hen by overhunting is a classic example of extinction, one of the best-documented in modern times. Human activities played the principal role, but several complexities of the case make it useful as an introduction to biological causes of extinction.

In colonial America, the heath hen was edible, easy to kill, and abundant over much of the eastern seaboard, from Maine to Virginia. Intensive hunting, coupled with habitat destruction by an expanding human population, gradually reduced the heath hens' geographic range. By 1840, they were limited to Long Island, parts of Pennsylvania, New Jersey, and a few other places. From 1870 on, they existed only on the island of Martha's Vineyard, off the coast of Massachusetts. The heath hen population there continued to decline until 1908, when a 1,600-acre refuge was established to protect the remaining fifty birds.

Following this protection, the population on Martha's Vineyard grew steadily. The birds spread over the entire island and numbered about two thousand by 1915. Hunting had long been forbidden, and the refuge was protected by fire control. So far, so good.

Then, starting in 1916, a series of mostly natural events led to the final extinction. These were (1) a natural fire, spread by a strong gale, that destroyed much of the breeding area; (2) a hard winter, immediately following the fire, and (coincidentally) accompanied by an unusual influx of predatory goshawks; (3) inbreeding, caused by the reduced population size and the accident of a distorted sex ratio; and (4) a poultry disease, introduced from domestic turkeys, that killed a substantial number of the remaining birds. By 1927, only eleven males and two females were left. By the end of 1928, only one bird remained. It was last seen on March 11, 1932.

The demise of the heath hen was not actually a species extinction. This bird was one of several subspecies (or varieties) of the species *Tympanuchus cupido,* best known now as the greater prairie chicken, occupying a fairly large range in the American Midwest and Plains states. Nevertheless, the case is relevant to the general problem of extinction.

The important thing about the heath hen extinction is that it developed in two distinct stages. The first was devastation by a new and sudden stress—human hunting. This served to reduce the geographic range drastically. The second stage, starting in 1916, was the series of accidents— some physical, some biological—that led to the final extinction. None of these accidents would have been significant if the range of the species had not already been limited to Martha's Vineyard. The fire, the goshawk predation, the inbreeding, and probably the poultry disease could not have done the job had the heath hen still ranged from Maine to Virginia. Even though some populations might have been eliminated by localized stresses, and probably were, the heath hens as a group would probably have continued to thrive had it not been for the hunting.

IMPORTANCE OF THE FIRST STRIKE

Can we generalize from the case of the heath hen? Does the extinction of a well-established species require a substantial initial hit—a first strike—to reduce geographic range, followed by a run of bad luck to finish the job? Maybe. There is just one problem: Could slow-acting stresses, like goshawk predation or increasing incidence of hard winters, have produced the extinction without a first strike? This gets into the problem of the long-term effects of slow processes and has an important bearing on extinction over geologic time spans.

In numerous writings on the heath hen, the consensus is that the subspecies would have been virtually immortal without that first strike. Simberloff has commented that "natural extinction on continents must be very rare." By "natural" he means without the influence of man and by "continents" he means species distributed over large areas. But how rare is rare? To a biologist, survival for hundreds or thousands of years is tantamount to immortality. For the paleontologist, a thousand years is trivial because a tiny disadvantage in the struggle for survival could be decisive if applied over millions of years. This was the essence of my conclusion about the failure of the trilobites. They had, for whatever reasons, a survival rate slightly lower than that of other marine animals. This led to their extinction even though it took 325 million years.

The foregoing still does not tell us whether a first strike is generally (or usually) necessary for the extinction of a well-established species. This remains an open question. We have established that the first strike, when present, greatly accelerates the process. And we have established (I think) that extinction can occur without a first strike. But we have not established which scenario is most common, and therefore most important, in the history of life.

PROBLEMS OF SMALL POPULATIONS

The risk of extinction in small populations has been well studied by ecologists engaged in a relatively new discipline called conservation biology. Because one of the aims of conservation biology is to devise ways of preserving endangered species, the problem of small populations has attracted special attention and scientific talent.

From this work has come the concept of the *minimum viable population* (MVP), an idea developed by Robert MacArthur and E. O. Wilson in 1967. In Dan Simberloff's words, "Populations above this point [are] virtually immune to extinction, while those below this point [are] likely to go extinct very quickly." Simberloff lists the four most common causes of extinction of populations below their MVP:

1. *Demographic stochasticity*. This is basically the Gambler's Ruin of Chapter 3. (The word *stochasticity* means, for all practical purposes, "randomness.") When populations are very small, minor glitches in mating, reproduction, or survival of young can cause the population size to drift downward to the absorbing boundary at zero. In other words, if populations are already very small, a modest run of bad luck can do them in.

2. *Genetic deterioration*. Small populations have, perforce, smaller genomes than do larger populations. This means, in extreme cases at least, that the species may

not have the genetic variability to adapt to changing conditions. Also, small populations are prone to what is called "genetic drift," wherein the genome may change in random directions regardless of—or despite—natural selection.

3. *Social dysfunction*. This broad category covers the deterioration that occurs in certain behavioral traits when populations become too small. For example, in species that are not gregarious, the maintainance of populations may depend on the ability of males and females to find each other for breeding. If the distribution of the population becomes too sparse, birthrates will drop.

4. *Extrinsic forces*. This covers the wide variety of large and small perturbations typified by fire, disease, and other problems that affected the final heath hen population on Martha's Vineyard. Whereas the first three categories in Simberloff's list are products of a population's smallness, extrinsic forces hit populations of any size but inflict serious harm only on smaller populations. If the heath hen had been able to maintain even small populations distant from Martha's Vineyard, it might have survived because many of the extrinsic forces were local. Simberloff has pointed out that for several of the extrinsic forces, the geographic extent of the species may be more important than the number of individuals.

How small must a population be for Simberloff's four factors to be significant? The size of the minimum viable population (MVP) has been studied over a wide range of conditions, with some interesting results. The MVP varies greatly from organism to organism. One of the most important variables is the inherent birthrate of the organism. With high birthrates, a species can rebound from difficulty quickly and also remain (during good times) close to the carrying capacity of its environment.

Despite variability in the MVP, all studies come to the same striking conclusion: MVP size is very low, commonly in the range of a few tens or hundreds of individuals. The heath hen, with two thousand birds in 1915, may have been close to its minimum viable population.

The risks imposed by small population size are real and important. They often provide the coup de grace that completes the extinction process. But for the well-established species that dominate the fossil record, small population size becomes significant only after a first strike—or its long-term equivalent of slow deterioration.

Let me note in passing something that many readers may have observed. The extinction theory developed by biologists is dominated by the analysis of organisms rather like us—mobile, terrestrial, bisexually reproducing animals. Why? In part because we are land-dwelling vertebrate animals ourselves and in part because large, terrestrial animals are easy to study. Much of our knowledge of contemporary extinction comes from birds. This is because of the vast accumulation of observations made over several centuries by dedicated amateurs.

The extinction problem may be quite different for other groups of organisms. What of plants or sponges or marine zooplankton? What, for example, is the MVP of an oyster? Oysters, like many marine invertebrates, have breeding systems in which all fertilization is external; nothing in an oyster is comparable to the mating behavior of birds and mammals. In a good (or lucky) year, a single oyster may produce tens of thousands of successful offspring; in a bad year, none. Because of these differences, population sizes in oysters and similar creatures tend to fluctuate wildly. This surely affects species already vulnerable to extinction, but the careful analyses done for birds and mammals have not yet been extended to most other parts of the plant and animal world.

COMPETITION

Considering Darwin's emphasis on competition, one might expect it to be first on any list of biological causes of extinction. The struggle for existence has long been so portrayed—competition between predators and their prey, among predators for prey, and among prey for survival. Competition is probably the single most common theme of TV nature programs.

Within the scientific community, ecologists and evolutionary biologists have also emphasized competition. It has seemed self-evident that competition is an influential factor. But increasingly, ecologists have downplayed its role. Competition exists, of course, but may not be as crucial as once thought, especially in extinction.

Some studies of modern species have been designed to find out whether competition lowers the chances of survival. The best studies deal with so-called land-bridge islands—islands like Trinidad and Tasmania once connected to an adjacent mainland but now isolated by postglacial sea level rise. Before separation, the island fauna and flora matched that of the adjacent mainland. But after the separation, extinctions of species on the island occurred, because it was not large enough to support the full complement of species. (The relationship between land area and number of species will be described in the next section.)

Land-bridge islands are ideal for studying extinction in the recent past, because it can be assumed that any species now found on the mainland but not on the island died out on the island after separation. This logic carries some risk, but when fairly large numbers of species are considered statistically, the results are probably robust. Jared Diamond and several other ecologists have done detailed studies of extinction on land-bridge islands, as well as on so-called habitat islands—small patches recently separated from other areas of the same habitat.

In a study of Trinidad as a land-bridge island, John Terborgh and Blair Winter of Princeton asked whether bird species in close competition with others on the mainland had a higher-than-average extinction rate on the island. That is, were species with close competitors on the mainland more likely to be missing from Trinidad? Because enough behavioral information to assess competition on a case-by-case basis is rare, Terborgh and Winter used a proxy. They assumed that bird species normally living with other species of their own genus (congeners) were likely to be in competition with those species. That assumption simplified their problem to that of counting victims and survivors and seeing whether species with congeners on the mainland were more likely to be missing from Trinidad. The answer to that question was no. As many species with congeners on the mainland survived on Trinidad as did species without.

These are negative results, of course, and are thus difficult to interpret. Terborgh and Winter found no relationship between competition and proneness to extinction, but this could be because of a variety of unseen complicating factors. One can rarely prove that something unseen is not there. After exhaustive analyses of many aspects of the natural history of these birds, however, Terborgh and Winter concluded with the following rather strong statement: "Extinction is impartial in choosing its victims; species of all sizes [meaning body size], trophic levels and taxonomic groups fall prey. Rarity proves to be the best index of vulnerability." And this takes us back to MVP and Gambler's Ruin. Species with small populations are the most likely to go extinct.

SPECIES–AREA EFFECTS

A cornerstone of conservation biology is the relationship between the size of an area and the number of species it can sustain. Figure 7-1 shows the typical relationship between area and number of species. The raw data are counts of species found in tracts at least partly isolated from each other. The areas may be true islands, surrounded by water, or they may be habitat islands. In one of Jared Diamond's studies, the islands are mountaintops in Nevada and adjacent states. Outlines of these mountain islands are defined by the 7,500-foot contour. For small mammals that normally cannot cross desert valleys, that contour is comparable to the shore of an oceanic island. The number of species found on each island correlates well with the area of the island.

If the species–area relationship were a straight line, it

Species–Area Effect

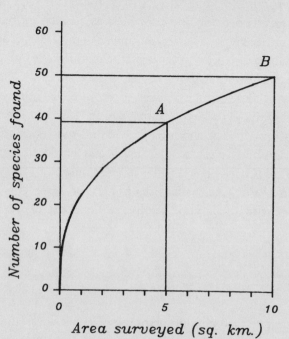

FIGURE 7-1. An example of the relationship between geographic area surveyed and number of species found. The general shape of the curve is typical, although its precise shape varies with the biology and natural history of the area being studied. In the case shown, a doubling of the area surveyed adds only about 25 percent to the number of species. The species-area relationship is used widely by conservation biologists in the design of parks and other protected areas and in predicting the number of extinctions that will result from reduction of habitable area.

would be of no great interest. But notice the effect of the curve. If we doubled the area at point A in Figure 7-1 (from five square kilometers to ten), we would increase the number of species (point B) but not double it—species would increase only from 39 to 50. Conversely, if we were to cut area in half (from B to A), the reduction in species would be less than half (from 50 to 39).

Suppose we have an island with 100 species, and we divide it into equal halves by erecting a tall fence across the middle. Suppose further that, at the time of the fence building, all 100 species lived over the entire island. Then we wait. If the species-area effect is working, neither half of the island will be able to support all 100 species. Extinctions will occur in both halves, reducing each to the number sustainable by the smaller area. Total biodiversity will have been reduced if some of the same species happened to die out in each half.

In conservation biology, the species-area effect is widely used to predict species loss with removal of habitat area. Depending on the shape of the species-area curve for a given region and plant or animal group, species loss can be minimized by a judicious control of the amount and distribution of habitat placed in reserves—that is, in artificial islands.

Recall Terborgh and Winter's experience with the extinction of birds on Trinidad; they found it to be nonselective (random) with respect to all characteristics except initial rarity—only population size predicted which species would to go extinct. In theory, therefore, one could design a refuge system to preserve a maximum number of species, with some species surviving in one refuge and others, by chance, in another refuge. But, as Terborgh and Winter also found, the species that were rare to begin with might not survive in either refuge.

The foregoing discussion barely scratches the surface of a complex and growing field of research. Although the species-area concept has been criticized—it is not as neat as I have presented it—the approach remains central to extinction-related conservation efforts.

SPECIES-AREA AND PAST EXTINCTIONS

The amount of habitable area available for life has changed through geologic time. In the marine realm, a lowering of sea level dries out vast areas of the shallow continental shelves, thus reducing the habitable area available to bottom-dwelling, marine species. The same lowering drains the broad inland seas that mark many periods in earth history, such as the Cretaceous. Conversely, a sea level rise creates new areas for marine organisms. Similar effects occur on land as the inverse: sea level lowering adds habitable area.

The rise and fall of land bridges between major continents is like the erection or removal of fences across large islands. The Isthmus of Panama, for example, has in the past few million years alternated several times between being emergent and being submergent. When Panama is up, land animals have a natural corridor for migration north and south but marine animals in the Gulf of Mexico are isolated from those of the eastern Pacific. When Panama is down, the reverse occurs.

THE GREAT AMERICAN INTERCHANGE

The species-area effects of the Panamanian land bridge have been documented especially well for land mammals. During much of mammalian evolution, the land bridge was submerged, isolating North from South America. The two mammal faunas developed quite differently; marsupials dominated in South America and placentals in North America. Some migration occurred back and forth: so-called waif immigrants were able to move between the continents by island-hopping up or down the antillean chain. But each continent had an essentially independent assemblage of mammals, and the number of species in each appears to have been in balance with the habitable area.

Then, about three million years ago, the land bridge slowly emerged and mammals literally walked from north to south and from south to north. North American mammals moving south included a number of species of skunks, peccaries, wolves, foxes, bears, camels, horses, tapirs, mastodons, and several other groups. Among those walking north were armadillos, porcupines, oppossums, tree and ground sloths, monkeys, and anteaters. The interchange was unequal: more North American natives moved south than South Americans north. This inequality stemmed directly from the fact that there were more North American species to start with—that is, before the interchange.

Species-area curves (such as the one in Figure 7-1) suggest that the combined land areas of North and South America could not support all the species that had previously occupied the separate continents. Given the migration back and forth, extinction was inevitable.

In both North and South America, the interchange first increased the number of species in each area. Then, as species-area limitations began to operate, the number of species in each area dropped back to levels somewhat lower than those before the interchange. Today, about 50 percent of the South American mammal genera are of North American origin, and about 20 percent of the North American genera came from South America. Because of the extinctions produced by the interchange, and because of Pleistocene extinctions on both continents, the total fauna is smaller than before.

THE HISTORY OF TROPICAL RAIN FORESTS

We are accustomed to thinking of tropical rain forests as the stable product of slow evolution over hundreds of millions of years. They are not! The geologic record indicates that rain forests have occurred only spottily in space and time, because they require an unusual set of circumstances, which have existed for only relatively short periods.

The rain forests we have now in the Amazon Basin, West Africa, and elsewhere depend on relatively *low* global temperatures to reduce seasonality in equatorial regions, on a configuration of continents and terrain conducive to heavy rainfall in the tropics, and on substantial time for the species making up complex communities to evolve. Fred Ziegler, a paleogeographer and climatologist at the University of Chicago, has estimated that, during the 350 million years since diverse land floras first evolved, tropical rain forests have flourished for only about a quarter of the time.

The shifts from rain forest to non—rain forest must have caused massive changes in habitable area and surely triggered species extinctions. The shifts may also have had a constructive effect: each time rain forests reevolved from scattered survivors, there was opportunity for evolutionary innovation, including new adaptations.

An interesting sidelight on the rain forest phenomenon is its bearing on biodiversity. Today, a majority of plant and animal species live in the wet tropics. If today's rain forests were lost, global biodiversity would drop. Because of fluctuations in the rain forest presence in the geologic past, global biodiversity must also have fluctuated; it was probably both higher and lower than it is today.

The history of our present rain forests over the past fifty thousand years is of particular interest. While the historical record is slim—fossil preservation in the tropics is poor—enough is known to raise intriguing questions. There is considerable evidence that the Amazon Basin and West Africa became cooler and drier at least four times during the past fifty thousand years, and the rain forests must have contracted.

Figure 7-2 shows one estimate of the change in South America. The map on the left shows the present extent of tropical rain forest; the map on the right shows the probable refuges occupied by rain forest during the dry periods. Similar maps have been constructed for West Africa: they also show the breakup of a large, continuous area into many small patches. It should be added that the rain forest maps for the late Pleistocene are controversial, because the paucity of good documentation necessitates the use of sometimes questionable proxy data.

Tropical Rain Forest Distributions

Present day   Pleistocene dry periods
(inferred)

FIGURE 7-2. Comparison of present rain forest distribution in South America with that inferred for glacial intervals when the region had a drier climate. The inferred pattern is based on circumstantial evidence; more geologic research is needed to confirm the pattern. (Redrawn from Simberloff, 1986.)

Quite apart from the obvious ones for conservation biology, the history of the rain forests is pregnant with broader evolutionary implications. Suppose, for example, that we accept the proposition that rain forest area was reduced by 84 percent in the very recent past, as is indicated by Figure 7-2. Many extinctions would have occurred, some through the species-area effect and some because of the loss of species naturally confined to areas that lost forest. Remember that many tropical species, particularly insects, are found only in one small area, perhaps a single tree.

If we accept the foregoing, we are faced with the problem of explaining how a substantial portion of the tremendous diversity of the present rain forests evolved in less than fifty thousand years. This seems an incredible speciation rate and raises more questions than it answers.

The thin factual basis of the recent history of rain forests can be remedied, but only through intensive research programs that work with climatic history and paleontological evidence. Until this is done, the history of rain forests, and its evolutionary implications, will remain enigmatic.

SOURCES AND FURTHER READING

Halliday, T. 1978. *Vanishing Birds: Their natural history and conservation.* New York: Holt, Rinehart, and Winston. Contains information on the case of the heath hen.

MacArthur, R. H., and E. O. Wilson. 1967. *The theory of island biogeography.* Princeton: Princeton University Press. A classic monograph developing the species-area idea.

Marshall, L. G., S. D. Webb, J. J. Sepkoski, Jr., and D. M. Raup. 1982. Mammalian evolution and the Great American Interchange. *Science* 215:1351–57.

Simberloff, D. 1986. Are we on the verge of a mass extinction in tropical rain forests? In *Dynamics of extinction,* ed. D. K. Elliott, 165–80. New York: Wiley-Interscience. Contains discussion of the postglacial history of rain forests; source for Figure 7-2.

Soule, M. E., ed. 1986. *Conservation biology: The science of scarcity and diversity.* Sunderland, Mass.: Sinauer Associates. An excellent collection of research articles on conservation biology.

Terborgh, J., and B. Winter. 1980. Some causes of extinction. In *Conservation biology: An evolutionary-ecological perspective,* ed. M. E. Soule and B. A. Wilcox, 119–33. Sunderland, Mass.: Sinauer Associates.

Ziegler, A. M., et al. 1987. Coal, climate and terrestrial productivity: The present and early Cretaceous compared. In *Coal and coal-bearing strata: Recent advances,* ed. A. C. Scott. Geological Society of America, Special Publication 32. Pp. 25–49. Boulder: GSA. A research article analyzing some aspects of the history of climate, especially those relevant to rain forests.

CHAPTER 8

PHYSICAL CAUSES OF EXTINCTION

In a college textbook titled *History of Life* (1976), Richard Cowan of the University of California (Davis) wrote the following on causes of the Cretaceous mass extinction:

> All kinds of explanations have been offered (impacting meteorites, huge volcanic eruptions, bursts of deadly radiation from the sun, selenium poisoning from volcanic ash, Noah's Flood, and so on). But as sensible scientists we should try to find a fairly ordinary set of circumstances to explain our facts before resorting to explanations that would be very difficult to prove or disprove.

This passage captures the philosophy that has long guided geologists and paleontologists: the present is the key to the past. Impacting meteorites, huge volcanic eruptions, bursts of deadly radiation, and the like have never been recorded in human history (the "present") at intensities necessary to cause a mass extinction and thus can be ruled out.

In *History of Life,* Cowan goes on to discuss several "sensible" and "fairly ordinary" causes of the mass extinction. He ends with this statement:

> So essentially we can conclude that climatic change, induced in the first place by continental movements, played a major part in causing the biological changes at the end of the Cretaceous. Obviously this story is still rather skeletal, and needs to be fleshed out with solid facts. But it provides the best explanation yet for the Cretaceous extinction which followed so closely after the Middle Cretaceous continental flooding.

In the pages between the quoted passages, Cowan presents brief but reasonable justifications for his conclusion. He notes that, because the extinctions include both terrestrial and marine organisms, a "truly worldwide" force is needed, thus ruling out several candidates.

He chooses climatic change as the culprit because it could plausibly affect all environments and because there is geologic evidence of global cooling near the end of Cretaceous time. To this, Cowan adds selective extinction of large animals, noting that they often "live dangerously close in an ecological sense." Cowan is not claiming that the case is settled—only that climate change is the most likely explanation.

TRADITIONAL FAVORITES

Traditionally, physical explanations of extinction come in two sizes: one for mass extinction and another for background extinction. For mass extinction, changes in global climate and sea level are by far the most popular. Another is change in the salinity of ocean water. Yet another is anoxia, the depletion of oxygen in shallow, marine environments.

For the smaller extinctions, a host of physical causes has been proposed, such as regional climatic change on land with accompanying changes in rates of erosion and therefore in the amount and nature of sediments transported to the oceans. Generally, however, small extinction events have received little attention. Perhaps they are viewed as being so complex or so inevitable that a search for causes is futile—even unnecessary.

SEA LEVEL AND CLIMATE

Sea level has fluctuated as long as there have been oceans. It is relatively low today, although during peak periods of Pleistocene glaciation, it was two hundred meters lower. In small areas, sea level is influenced by local movements of the earth's crust. The land surface is pushed up or down by tectonic forces, shifting the intersection of land and sea. Sea level is even affected by human activities, as in the subsidence caused by years of petroleum removal from rocks underlying Los Angeles.

Larger changes of sea level are caused either by glaciation

or by crustal shifts that change the shape of ocean basins. Over the past several centuries, global sea level has risen as polar ice melted. However, this has been offset in some places by local movements, some also stemming from glaciation. Over much of Scandinavia, for example, sea level has gone down steadily because the crust is still rebounding following the removal of the former ice sheet.

Monitoring changes in sea level through earth history is straightforward in theory but not in practice. Fossils often indicate the distribution of land and sea: organisms known to be terrestrial indicate deposition on continents, and marine organisms indicate deposition on ocean bottoms. Physical evidence from sedimentary rocks provides information, too, especially in the distinctive features of shallow-water and beach deposits. A common difficulty, however, is that local or regional effects can blur the global record. Figure 8-1 shows an example of a sea level curve—the so-called Vail curve, developed by an Exxon research group.

A nontrivial problem with sea level curves is that they are sometimes secret. Petroleum companies invest large sums of R & D money to track ancient sea level, because oil and gas tend to accumulate near old shorelines. The curve shown in Figure 8-1 was published in the open scientific literature, but such results are often withheld—or released with some

FIGURE 8-1.   One estimate of the Phanerozoic history of global sea level. Some of the larger fluctuations during the Pleistocene and other glaciations are not included. Note that sea level is at present somewhat lower than it was during most times in the past. (Adapted from Vail et al., 1977.)

Phanerozoic Sea Level Curve

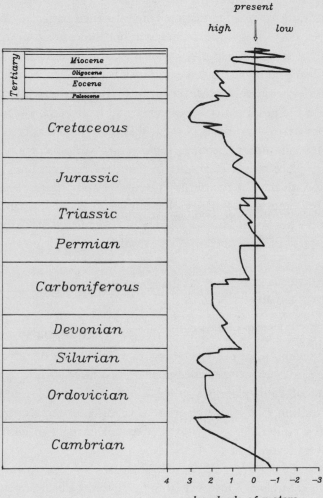

parts left blank. Sometimes, the full curve is released but the information used to construct it is not.

Chronologies of climatic change are even harder to construct. Climate is complex, being a conglomerate of temperature, circulation patterns, seasonality, and other factors. Of the many components, temperature is by far the best studied. Data for past temperatures come from distributions of fossils known to be climatically restricted and from certain chemical analyses—mostly isotopic—that pick up actual records of temperature. Of course, temperature varies enormously from place to place; no single temperature recording (on land or in the sea) is typical. This further complicates the job of building a chronology. Despite these problems, curves for the history of climate have been constructed. An example based on temperature is shown in Figure 8-2.

SPECIES-AREA EFFECTS

As we already noted, changes in the distribution of land and sea influence the habitable area available for organisms. We saw how the species-area effect produced some extinctions of land mammals when North and South America were connected by the Isthmus of Panama. In this section, I will consider the species-area effects of changes in climate and sea level.

Consider marine life of the broad shelves that surround most continents. Outward from the shore of a typical continent, the bottom deepens gently to a depth of about 135 meters (global average). The inclination of the shelf is slight, averaging barely more than a tenth of a degree. The shelf edge is marked by a change in angle, where the sea

Ocean Water Temperature
(northwest Europe)

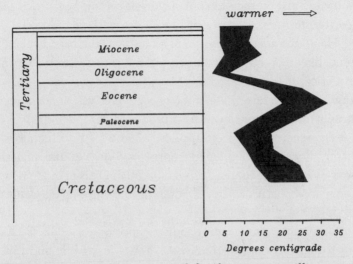

FIGURE 8-2.   Temperature record for the past 100 million years inferred from oxygen isotope ratios found in fossil shells of marine mollusks. The width of the band reflects variation among analyses. Although temperature fluctuations are large, the overall trend is one of cooling as the present day is approached. This trend is typical of records from other fossils in other regions, indicating that cooling was global. (Redrawn from Anderson, 1990.)

bottom drops off more sharply on what is called the continental slope. Continental shelves vary greatly in width, from almost zero off southern California to several hundred miles off eastern North America.

Continental shelves, high in nutrients, have always supported more biomass than the deep ocean has. Yet the shelves are well within reach of sea level fluctuations. If continental shelves are drained, the habitat for shelf dwelling species is sharply reduced. Relatively shallow areas continue to exist on the old continental slope, but because of its greater steepness, the area of favorable depths is reduced.

Habitable area has also been a factor in the broad inland seas, like that covering much of interior western North America in the Cretaceous. Many of these seas were shallow, perhaps a few tens of meters deep over vast areas. Thus, only a moderate sea level drop produced complete drying.

The species-area effect applies at some scale to the alternately flooded and drained continental shelves and inland seas. When flooded, the shelves and inland seas provide new areas for speciation, but drying of the same areas causes extinction.

Climatic change also produces species-area effects. As global temperature falls, climatic zones in middle and low latitudes become narrower. Isotherms (lines of equal temperature) are pushed southward in the Northern Hemisphere and northward in the Southern Hemisphere. If global temperature increases, isotherms move away from the equator, making tropical zones wider. This bellows-like expansion and contraction of climatic zones increases and decreases habitable area for many plants and animals.

In view of all this, it is not surprising that changes in climate and sea level—operating through species-area effects—are prime candidates to serve as explanations for past extinctions.

TESTING SEA LEVEL AND CLIMATE

Almost no attempt has ever been made to test statistically the correspondence between extinction and sea level or temperature. This is partly because of the uncertainties, already mentioned, in sea level and temperature data. But it is also due, I think, to the fact that most geologists and paleontologists have little training in the appropriate methods. The extinction problem calls for techniques from what is known as *time series analysis,* a branch of statistics missing from most basic courses.

Therefore, the arguments for climate and sea level are usually presented rather qualitatively as accumulations of case studies, often including one or more of the Big Five mass extinctions. I don't mean to suggest that a nonmathematical approach is necessarily bad. Science would be in difficult straits if only theories backed by rigorous mathematics were acceptable. But the case study approach makes it difficult to compare and evaluate the arguments.

An interesting test of the strength of the species-area effect of sea level lowering was carried out by my Chicago colleague Dave Jablonski, working with Karl Flessa of the University of Arizona. They asked an importation question: If all life on present-day continental shelves were to be eliminated, how many global extinctions would be recorded? Jablonski and Flessa assembled data on geography of a wide variety of living marine mollusks. They then tabulated the global fauna that would be left after the elimination of continental shelves and calculated a hypothetical extinction rate. This is an extreme scenario because it does not allow shelf species to escape by moving offshore, and therefore the extinction estimate Jablonski and Flessa produced is an upper limit, a worst case.

They found an astonishingly low extinction rate of only 13 percent of all families, far lower than the rate (52 percent) for families of comparable animals in the Permian mass extinction. The explanation is straightforward: 87 percent of all living families have at least one species in shallow water around steep-sided oceanic islands—islands where the coastline would move outward during a regression, but without loss of habitable area.

It could be argued in response to Jablonski and Flessa that draining continental shelves is merely a first strike that reduces population sizes and geographic ranges, so the survivors on oceanic islands are susceptible to the coup de grace of lesser stresses. However, Jablonski and Flessa have countered this by noting that most of the 87 percent of the families with island refuges are found on two or more islands.

The work of Jablonski and Flessa casts doubt on the efficacy of marine regression—whether caused by glaciation or tectonics—as a cause of mass extinction, but I don't think it closes the door on it. Much more hard-nosed testing is needed.

There is not space here to give a full review of arguments for and against climate and sea level as the principal cause(s) of major extinctions. The arguments are complex and call on detailed analyses of particular extinction events. I am obliged, therefore, to leave the reader only with my opinion that neither climate nor sea level has been shown convincingly to have caused any major extinction events. However,

I do have one more solid reason for being skeptical: the Pleistocene glaciation.

THE PLEISTOCENE EXPERIENCE

If a mass extinction can be produced by a change in sea level or climate, the recent continental glaciation should have produced a beaut!

The Pleistocene epoch, as it is officially called, started 1.64 million years ago. It ended 10,000 years ago, with the last retreat of continental ice sheets. Stopping the Pleistocene at 10,000 yr BP is an optimistic judgment, for it assumes that ice sheets have retreated for the last time. Because the glaciation came in several pulses separated by mild intervals, we could be in an interglacial period now. Maybe the Pleistocene isn't over.

The Pleistocene record is well known. It happened only yesterday geologically and left deposits readily studied on land surfaces. It was a time of major alteration of global environments affecting terrestrial as well as marine settings. Sea level went up and down hundreds of meters, and temperature zones were compressed toward the equator. Major weather patterns, especially the monsoon systems, were markedly different from today's. There were also pronounced changes in atmospheric carbon dioxide.

Sea level fluctuations altered geography, with islands and isthmuses forming and disappearing as the glaciers waxed and waned. Although earth history has seen many glacial epochs, the Pleistocene was probably among the most se-

149

vere; the exact ranking is difficult to estimate because the older ones are not as well known.

The Pleistocene experienced some extinctions but nothing like the Big Five. The intensity was much less than for those events used to define periods and lesser units of geologic time. Except for the large land mammals already discussed in the context of the blitzkrieg theory, Pleistocene extinctions were spotty and idiosyncratic. A few biological groups in a few places suffered, but only a few genera or families were lost. Perhaps the closest to a mass extinction occurred among marine mollusks of the tropical western Atlantic and Caribbean, starting with preglacial cooling about 3.5 ma BP. This case has been well studied by Steven Stanley, who concludes that it was a "regional mass extinction" of faunas caught geographically by the temperature drop. Stanley's parallel analysis of Pacific mollusks showed no significant extinction.

So, although extinctions occurred during the Pleistocene, some of them related to climate or sea level, they were not in a league with the mass extinctions. By the same token, there is no significant association between extinction and glaciation further back in time. The Cretaceous mass extinction, for example, occurred during a long ice-free interval. The Pleistocene experience, therefore, can only dampen one's enthusiasm for temperature and sea level as general explanations of extinction.

At this point, I can almost hear some of my colleagues complaining that I am pressing too hard to find a single cause of extinction. Why not accept that most extinction events are complex—and different from each other? Why not have marine regression dominating in the Permian, climate in the Cretaceous, and something totally different in the Devonian?

My response is that this may in fact be the way the world works, but I hope not, because it would be hard to prove, for the following reason. Suppose major regression and mass extinction each occurred only once in earth history. If these two unique events coincided in time, we would have a strong case for regression as the cause of the extinction. But regressions and mass extinctions are scattered throughout geologic history, along with a lot of other things. There is no way of assessing cause and effect except to look for patterns of coincidence—and this requires multiple examples of each cause-and-effect pair. If all extinction events are different, the deciphering of any one of them will be next to impossible.

Exotic Physical Causes

If "normal" geologic processes lack power to cause significant extinction (as I assert), what are the alternatives? By definition, they are processes (or phenomena) we have never experienced. It is a good thing that our experiences have been gentler, but this forces us into a speculative mode.

Any exotic cause needs two supporting facts—that it happened and that it happened in conjunction with extinction. The process or phenomenon must leave unequivocal traces, a smoking gun. By far the best candidate so far is the collision of earth with solar system debris: comets and asteroids. Because this case is rather strong, I will devote the next chapter to it. But first, some other possibilities.

UNHEARD-OF VOLCANISM

One conceivable agent of extinction is volcanism more intense than any in human history. A suggested scenario calls for explosive volcanism like the Krakatoa eruption of 1883 but on a grander scale—with many large volcanoes going off at the same time. Krakatoa started with an explosion equivalent to a hundred megatons of TNT. It put so much debris, ash, and sulfate aerosols into the atmosphere that the effects were noticeable for several years. The global temperature dropped by a few degrees because of the shielding of solar radiation. Despite this power, the Krakatoa eruption had no known biological effects outside the immediate area. But a thousand Krakatoas acting simultaneously might trigger a large climatic change, perhaps even cutting out enough sunlight to interfere with photosynthesis.

Volcanologists are quick to point out that there is no evidence that enough big volcanoes ever explode at the same time to do the job, nor is there any known mechanism for causing coordinated eruptions. Moreover, volcanic activity usually leaves a durable record, and there is no geologic evidence for bursts of simultaneous eruptions. On the other hand, current methods of geologic dating may not be good enough to establish the critical evidence of simultaneity. The late Cretaceous was indeed a time of volcanic activity, but evidence of simultaneity is lacking.

Another kind of volcanic activity—flood-type volcanism—produces broad, thick accumulations of lava without explosive eruption. The volcanic rocks of the Columbia River and Snake River areas of the northwestern United

States are good examples. Another is the Deccan, a thick basaltic deposit that blankets about a third of India. Estimates of flow rates suggest that heat production was great; some climatologists have speculated that such a heat source could alter global circulation, especially if it was near the equator.

There are approximately six really large deposits of flood volcanism in the geologic record, including the Columbia River basalts and the Deccan. There is a hint of an age correspondence with mass extinction in several cases. For example, Deccan volcanism may have started at almost the time of the Cretaceous extinction. Unfortunately, other examples are so few and the dating uncertainties so great that good statistical testing is impossible. Thus, the proponents of flood volcanism (and there are a good number of them) have had much difficulty convincing their colleagues.

Cosmic Causes

Science fiction writers are fond of noting that outer space can be unpleasant. Stars blow up and run into each other. Planetary systems pass through giant molecular clouds. The cosmic environment is subject to all manner of bursts of high-energy radiation. Could such happenings have seriously affected life on earth? The history of life spans 15 to 30 percent of the history of the universe. A lot could have happened in that time.

A few attempts have been made to search for extinction mechanisms in space, apart from comets and asteroids. Possibilities range from the effects of nearby exploding stars (supernovae) to changes in the sun's radiation. None has been fruitful. The incidence of supernovae is reasonably well known, at least statistically, but the consensus among astronomers is that there have been too few close encounters for them to have done significant biological damage.

Variation in the sun's intensity (solar constant) is a promising candidate, but we have no records before the last few centuries. Some changes in the sun can be predicted from theories of stellar evolution, but these are long-term trends (such as the presumed 30 percent increase in the solar constant over earth history) rather than the short-lived phenomena needed for mass extinction.

I leave the subject of exotic causes with a sense of frustration. I have tried to evaluate agents of extinction that nobody has ever seen, thus inviting speculation and flights of imagination generally reserved for the lunatic fringe. But the alternative—the assumption that we know all the good candidates—is unacceptable.

SOURCES AND FURTHER READING

Anderson, T. F. 1990. Temperature from oxygen isotope ratios. In *Paleobiology, a synthesis,* ed. D. E. G. Briggs and P. R. Crowther, 403–6. Oxford: Blackwell. The source for Figure 8-2.

Cowan, R. 1976. *History of life.* New York: McGraw-Hill. A basic college text on paleontology; a new edition is in preparation.

Jablonski, D., and K. W. Flessa. 1984. The taxonomic structure of shallow-water marine faunas: Implications for Phanerozoic extinctions. *Malacologia* 27:43–66. A research paper testing the hypothesis that sea level lowering should cause extinction in marine organisms because of the species-area effect; recently updated by more data.

Milne, D. H., D. M. Raup, J. Billingham, K. Niklas, and K. Padian, eds. 1985. *The evolution of complex and higher organisms.* NASA Special Publication SP-478. Washington, D.C.: NASA. The report of a series of workshops held by NASA to evaluate extraterrestrial influences on evolution.

Vail, P. R., R. M. Mitchum, and S. Thompson. 1977. Global cycles of relative changes of sea level. In *Seismic stratigraphy— Applications to hydrocarbon exploration,* ed. C. E. Payton. American Association of Petroleum Geologists, Memoir 26. Pp. 83–97. Tulsa: AAPG. The source for Figure 8-1.

CHAPTER 9

ROCKS FALLING OUT OF THE SKY

A meteorite is a piece of asteroid or a rocky comet fragment that has fallen out of the sky. In college and graduate school in the 1950s, I learned a little about meteorites, but not very much. Their study has always been a small subdiscipline. For meteoriticists and cosmochemists, however, these exotic rocks provide a unique glimpse of the early solar system.

Meteorite studies are based on the few objects found soon enough after impact to escape weathering. Usually a few inches to a few feet across, they are often found accidentally in farmers' fields. Most are too small to make noticeable craters on impact. The largest known meteorite, still in place in Namibia, is estimated to weigh 66 tons, and the largest one on display is a 34-ton specimen at Hayden Planetarium, in New York City.

Part of the scientific catechism I learned in school was that during the solar system's early development, impacts by large objects were common and contributed importantly to the growth of earth and other terrestrial planets. But bombardment ended, it was said, well before earth's present surface was formed. It ended because the reservoir of debris left over from the early solar system had been exhausted. Craters could still be seen on the moon and on Mars, dating from the early bombardment, but those on earth had been eliminated by erosion. The small meteorites found on earth were but trivial remnants of the exhausted debris cloud. So went the story. Like so much of what we think we know, the story turned out to be only partly true.

The biggest mistake was missing the fact that large objects continued to fall after the early bombardment. One of the best-known examples is Meteor Crater, in Arizona, also called Barringer Crater. Meteor Crater was mentioned in textbooks of the 1950s but very grudgingly. Debate had raged for years about its origin and about that of every other circular structure on earth for which an impact origin had been claimed. Reading the scientific literature of the fifties, one gets an impression that geologists did not want to recognize impact craters of recent date. Most geologists were convinced that Meteor Crater was formed by volcanic eruption, even though the area is littered with metal fragments of nickel-iron alloy.

In the 1960s and 1970s, everything changed because of several exciting discoveries: (1) two forms of the mineral quartz—stishovite and coesite—were found to be unequivocal indicators of the high pressures produced by large impacts; (2) satellite photographs of earth revealed structures so craterlike that old excuses (volcanic calderas, accidents of erosion, and so on) fell away; (3) the dating of some lunar rocks returned by Apollo astronauts indicated craters much younger than the early bombardment; and (4) astronomical observations of near-earth space proved that there are still many large objects with orbits overlapping the orbit of earth—that is, asteroids that could collide with earth.

With this encouragement, geologists searched worldwide for impact craters. More than a hundred craters have been verified so far. In fact, Meteor Crater has been relegated to a minor role as one of the smallest (1.2 kilometers in diameter) and youngest (50,000 yr BP).

But most paleontologists and many geologists did not hear about the new work on craters, nor were they aware that the old view of early bombardment had collapsed. And this explains the horror and disbelief when the 1980 Alvarez paper proposed meteorite impact as the cause of the K–T mass extinction. It was like suggesting that the dinosaurs had been shot by little green men from a spaceship.

CRATERING RATES

The current consensus is that although there was indeed an early phase of heavy bombardment, followed by a steep decline, the bombardment has continued at a moderate rate until the present day. There is even evidence that the rate has increased over the past 600 million years.

The most recent estimates for the current cratering rate are given below. The numbers come from the work of Eugene Shoemaker, of the U.S. Geological Survey, who has made great contributions to cratering studies and, separately, to the telescopic search for asteroids in earth-crossing orbits. The estimates are expressed as average time intervals (waiting times) between craters of specified sizes. Included are craters made by asteroids (70 percent) and by comet nuclei (30 percent). Crater sizes can be converted, roughly at least, to the sizes of the objects making them. Depending on the speed, angle of impact, and other factors, a 1-kilometer object produces a 10–20-kilometer crater; a 10-kilometer object makes a 150-kilometer crater. Note that the table ignores the smallest craters like Meteor Crater—a pip-squeak, despite the impressive photographs and television pictures. Meteor Crater was made by an object with a diameter about the width of a football field. Like many natural phenomena I have discussed, cratering has a highly skewed distribution: many small events and few large ones.

For the extinction problem, we need to know how good Shoemaker's estimates are and whether the impacts have enough destructive power to account for species extinctions. Shoemaker states that his uncertainty is "at least a factor of 2." This means that the estimated average spacing of 110,000 years for impacts causing 10-kilometer craters (or larger) may be anywhere between 55,000 and 220,000 years—that

Crater Diameter	Average Time Interval
> 10 km	110,000 years
> 20 km	400,000 years
> 30 km	1.2 million years
> 50 km	12.5 million years
> 100 km	50 million years
> 150 km	100 million years

is, it could be half or double the preferred estimate. This range is large, to be sure, but it gets us into the ballpark. The uncertainty is rather well known, because it has been deduced independently from studies of craters on earth, craters on the moon and the planets, and from sightings of contemporary asteroids and comets.

If Shoemaker's preferred estimates are off, they are probably on the low side because craters are still being discovered and new asteroids sighted at an increasing rate. Furthermore, impact structures on earth are almost certainly undercounted, owing to oil company policy. Many craters are buried by younger rocks, and fractured rock in their vicinity provides an ideal medium for accumulation of oil and gas. The petroleum companies, newly aware of the value of subsurface craters, are loath to make the locations public. This is fortunately not a universal practice; a few years ago, oil companies operating on Nova Scotia's continental shelf found and released information on a buried crater important to the extinction problem: the Montagnais Crater, with a diameter of 45 kilometers. Canadian law required the release of this information.

DESTRUCTIVE POWER

Here, present-day observations tell us nothing. We have not experienced, nor has our civilization recorded, any impact even close to the smallest in Shoemaker's table. The largest we have is the Tunguska event. This occurred in an uninhabited part of Siberia on June 30, 1908; although no crater was found, all trees over thousands of square miles were knocked down by the shock wave. The incoming object, exploded on hitting the atmosphere. The energy released was equivalent to about twelve megatons of TNT, roughly that of a very large hydrogen bomb. The event was observed (and heard) by the passengers and crew of a Trans-Siberian express train about 350 miles away, but not studied on the ground until years later.

Estimates of meteorite destructive power are based almost entirely on theory or computer simulations based on theory. For the impact of even a one-kilometer object (ten football fields), the estimated energy release is truly incredible, exceeding by a large multiple that released by the simultaneous detonation of every nuclear weapon in existence. Luis Alvarez quoted one estimate for the impact of a 10-kilometer object: one hundred million megatons of TNT. Even if this number is too high, the destructive power defies the imagination.

The predicted effects of large impacts include shock waves, tsunamis (tidal waves), acid rain, forest fires, darkness caused by atmospheric dust and soot, and global heating or global cooling. The uncertainty about temperature stems from our ignorance of whether the debris-clogged atmosphere would cause greenhouse warming (due to trapped heat) or severe cold (due to reduced sunlight).

Except for these generalities, our knowledge of environmental effects is poor. Short of experiencing an impact ourselves, our best bet lies with the fossil record. Given well-dated impacts, what kinds of species living in what habitats were killed (if any)? If this approach works, it is another example of the past as a key to the present (and future).

ALVAREZ AND THE K–T EXTINCTION

The story of how the late Luis Alvarez, Nobel physicist and scientific gadfly, worked with his geologist son Walter and two chemists, Frank Asaro and Helen Michel, on the K–T mass extinction has been told many times. Readers wishing a full account should read the *National Geographic* article (June 1989) or any of several books devoted to the story—including, I suppose, my own effort, called *The Nemesis Affair* (1986). My treatment here will be brief.

The Alvarez group discovered high concentrations of the element iridium in rocks at the Cretaceous–Tertiary boundary in Italy, Denmark, and New Zealand. Whereas iridium is common in some meteorites, it is vanishingly rare in the earth's crust. The Alvarez group quite logically concluded that the K–T deposit was formed of debris from the impact of an asteroid or rocky comet. So much iridium was present that the impacting object (often called a bolide) must have had a diameter of ten kilometers. Because of the coincidence in timing with a big mass extinction, they proposed that the environmental effects of the impact caused the extinction. They attributed species kill primarily to a shutdown of photosynthesis, on land and in the sea, caused by the debris cloud.

The stormy reaction to the Alvarez publication stimulated so much new research that we now know far more about the K–T boundary and about mass extinction than ever before. Iridium concentrations have been found in scores of K–T boundary rocks around the world. More important, new indicators of impact have been discovered—shock-metamorphosed minerals, and isotopic signals. Even the mineral stishovite, indicating ultra-high pressure, has been found. Thus, iridium is now only one of many indicators (not even the best) of impact at the K–T.

The case for a K–T impact would benefit from a verified crater of the right size and age. This is not crucial for most geologists, because so much of the planet's surface lies under the oceans, where craters are hard to recognize, and because much of the Cretaceous seafloor has been destroyed (subducted) by plate tectonics. Most paleontologists, however, want a crater—a smoking gun—and until one is found, the Alvarez idea will remain controversial.

The problem of the missing crater may be on the verge of solution. As I write this chapter in June of 1990, the scientific community is digesting two recent reports focusing on the Caribbean. One investigates a possible crater underlying Yucatan, and the other describes rocks in Haiti that suggest deposition following a huge impact. In a commentary in the journal *Science* introducing one of these reports, Richard Kerr (scientist and staff writer) notes, "A credible impact site would complete the evidence for the theory that the mass extinction at the end of the Cretaceous period was triggered by a collision between Earth and an asteroid." Later in the same paragraph, Kerr states that there is "sufficient evidence for most of the scientists involved in the research." Kerr's commentary gives me a strong sense that the tide of opinion is turning. Perhaps within the next few months, it will be difficult to find anyone who ever doubted the impact-extinction link. This happened in the 1960s with the acceptance of plate tectonics and continental drift.

PERIODICITY OF EXTINCTION AND NEMESIS

In many people's minds, meteorite impact as a possible cause of the K–T extinction is inextricably tied to another problem—the claimed 26-million-year periodicity of major extinctions. But the two propositions are independent.

In 1984, Jack Sepkoski and I published a short paper presenting a statistical analysis of the principal extinctions over the past 250 million years, with the conclusion that their spacing is clocklike—occurring every 26 million years. This was not a new idea; in fact, we were only confirming a similar conclusion reached by A. G. Fischer and Michael Arthur (both then at Princeton) seven years earlier. Jack and I did not propose a causal mechanism for the periodicity, but we did suggest that it was probably extraterrestrial.

Accepting our conclusions, several astronomers proposed solar system or galactic mechanisms to explain periodicity on earth. The proposal receiving most attention was that our sun has a small companion star that, in the course of a 26-million-year orbit, sweeps close enough to our part of the solar system to bring comet showers down on earth. The companion star, given many names, is most often called Nemesis.

Much research and press attention has been devoted to periodicity and to the astronomers' explanations. Here is a brief summary of recent developments.

- Extinction data have been reanalyzed by a dozen statisticians, geologists, paleontologists, and astronomers.

The results are mixed: about half support the 26-million-year periodicity (with minor revision of the period in some cases), and about half find no convincing evidence for cycles of any duration.

- Most astronomers have rejected the companion star (Nemesis) and other proposed mechanisms for periodicity. The Nemesis idea failed because computer simulations indicated that a small companion star in a large orbit would be unstable—too easily perturbed by close encounters with other stars. Nemesis has not been detected despite considerable searching by an automated telescope.

My own view is that periodicity is alive and well as a description of extinction history during the past 250 million years—despite the lack of a viable mechanism. The debate has pretty much died down because most scientists involved have decided that the extinction data do not show periodicity. But the proposal is still on the table, awaiting new data or new ways of looking at old data.

Fortunately, periodicity is not vital to the subject of this book. From an evolutionary standpoint, it does not make much difference whether extinctions come at regular or irregular intervals. The kill curve, based on average waiting times between events, is virtually independent of periodicity.

SOURCES AND FURTHER READING

Alvarez, L. W. 1987. Mass extinctions caused by large bolide impacts. *Physics Today,* July, 24–33. A strong advocacy statement for impacts as a cause of extinction, by the principal originator of the idea.

Glen, W. 1990. What killed the dinosaurs? *American Scientist,* July–Aug., 354–70. An up-to-date review of the mass extinction controversy by a historian of science.

Goldsmith, D. 1985. *Nemesis: The death star and other theories of mass extinction.* New York: Walker. One of several popular treatments of mass extinction.

Hsu, K. J. 1988. *The great dying.* Orlando: Harcourt Brace Jovanovich. A readable treatment of mass extinctions, with emphasis on the influence of Charles Darwin on thinking about extinction and life in general.

Kerr, R. A. 1990. Dinosaur's death blow in the Caribbean Sea? *Science* 248:815. A commentary on the discovery of evidence of impact in the Caribbean Basin.

Muller, R. 1988. *Nemesis: The death star.* New York: Weidenfeld & Nicolson. A popular account of the search for causes of the K–T extinction by one of the originators of the Nemesis theory.

Raup, D. M. 1985. *The Nemesis affair: A story of the death of dinosaurs and the ways of science.* New York: W. W. Norton.

Silver, L. T., and P. H. Schultz, eds. 1982. *Geological implications of impacts of large asteroids and comets on the earth.* Geological Society of America, Special Paper 190. Boulder: GSA. An important collection of research papers based on a 1981 meeting at Snowbird, Utah.

CHAPTER 10

COULD ALL EXTINCTION BE CAUSED BY METEORITE IMPACT?

I suggested in the previous chapter that the decade-long argument over the impact theory of the K–T extinction may soon be over. Supporters of the theory have declared victory before, however, and the tide of opinion may well recede. Science is like that. Rather than continue the argument here, I will explore a broader question: Could meteorite impact be *the* primary cause of species extinctions in the history of life? Strangely, this is an easier question. Instead of arguing about a single event, where coincidence can play a large role, this chapter will consider the entire history of extinction and some of its patterns.

PLAUSIBILITY ARGUMENTS

Several times in the past couple of years, I have suggested to colleagues that meteorite impact might cause most extinctions. Reactions have been curious. The common response is something like "But couldn't some mass extinctions be caused by other factors?" When I said "extinctions," my listeners heard "mass extinctions." Impact as a cause of mass extinctions is seen as an issue to be debated, but impact as a cause of the majority of all extinctions is so bizarre that the words are not heard. At a 1988 conference on extinction, I presented a paper suggesting the universality of extinction by impact. The idea was apparently well received but largely because I labeled it a "thought experiment" and did not claim actually to believe it.

Is it reasonable to think that impacts of comets and asteroids could account for background extinction as well as for one or more of the Big Five? Yes, I think it is reasonable. Moving from chapter to chapter writing this book, I have been impressed by nature's great difficulty in eliminating species that cover large areas. The heath hen was, until human hunting, virtually safe—at least on human time scales. In such cases, a first strike to reduce geographic range is common and may even be necessary.

Rapid and severe environmental stress during Pleistocene glaciation did not produce a noticeable pulse of extinction. Two likely explanations for this are (1) that the range of any well-established species contains refuges large enough to sustain the species, and (2) that many (most?) species can migrate faster than climatic zones or shorelines move. Evidently, the Pleistocene glaciation was neither fast nor severe enough to kill a significant number of species. What other normal geologic processes exceed this pace and severity?

Continental drift is one of the few possibilities, especially through species-area effects—as large land masses separate or coalesce. I don't think this will explain a significant fraction of the extinctions, however. In Chapter 7, I described the extinctions of mammals triggered by migration along the Isthmus of Panama. The extinctions were real but generally minor and without substantial consequences.

The kill curve (Figure 4-5) establishes the relationship between *extinction* and *time;* the Shoemaker estimates of cratering rate (Chapter 9) give the relationship between *meteorite impact* and *time.* Time being the common element in these relationships, we can make a giant leap by asserting (as a working hypothesis, thought experiment, hunch, or anything you like) that all species extinction is caused by meteorite impact. Having done this, we can write equations for both relationships, combine them, and then eliminate the common variable—time. We are left with a single equation relating impact and extinction. Figure 10-1 is a graph of this equation, showing the species kill produced by a given size of impact on the presumption that impact is the sole cause of extinction. The curve is dashed above crater diameters of 150 kilometers, because this is the limit of Shoemaker's estimates of cratering rate.

The next question is whether the impact-extinction curve of Figure 10-1 is credible when compared with other information on effects of impact. The Alvarez study estimated that the K–T impact should have left a crater about 150 kilometers in diameter, for which the curve indicates a species kill of about 70 percent, close to the estimate of species extinction from the fossil record of that event. So far, so good. Above this level on the curve, species kill increases with crater diameter but very slowly. Because this part of the curve is dashed, its position may be wrong, but at least it does not predict complete annihilation of life. This gives some support to my original assertion because the mathematics could have turned out very differently, with the curve reaching a 100 percent kill at a crater diameter of, say, 200 kilometers.

At the low end of the impact-extinction curve, the killing effect of very small impacts (craters less than 10 kilometers across) approaches zero. This also fits our experience because many small craters are not associated with unusual extinctions. Taking the curve (and its equation) literally, we see that a 5 percent species extinction occurs with craters of 24.5 kilometers, and these occur, on average, every one million years. Five percent is roughly the extinction level that normally defines the "biostratigraphic zone"—the smallest unit in geologic time recognizable by fossils on a global or near-global basis. In many parts of the geologic column, paleontologists have estimated the average duration of a stratigraphic zone to be about one million years.

The foregoing comments do not in any sense prove the impact-extinction curve valid as a statement of cause and effect, but they do establish plausibility. The curve unites two relationships developed from completely independent sources—craters and fossils—yet the results are reasonable. This, in turn, adds strength to the assertion that impacts are *the* general cause of extinction.

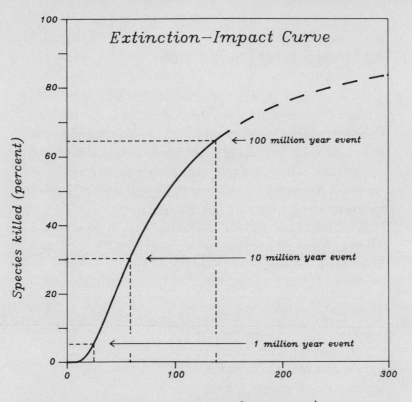

FIGURE 10-1. Species extinction in relation to size of crater formed by comet or asteroid impact. The curve is the inevitable result of combining the kill curve with Shoemaker's estimates of the rate of crater formation—on the working hypothesis that impact is primarily responsible for extinction. The degree to which this curve is credible is a measure of the validity of the hypothesis.

ARGUMENTS FROM OBSERVATION

Next we must consider whether there really is a correspondence between impact dates and extinction. I will try two approaches. One is to look at the abundant iridium data published in the past decade. There are better indicators of impact than iridium, but this feature has been surveyed most widely. The second approach is to look at ages of known craters to see whether they match extinctions.

Well-confirmed iridium anomalies have been reported from six geologic horizons other than the K–T boundary. These are as follows:

Geologic Stage	Age of Iridium Anomaly (ma BP)
Pliocene (Tertiary)	3
Middle Miocene (Tertiary)	12
Eocene (Tertiary)	35
[K–T boundary 65]	
Cenomanian (Cretaceous)	90
Callovian (Jurassic)	157
Frasnian (Devonian)	367

Whether these six anomalies support the impact-extinction link is debatable and depends on many factors. The oldest iridium anomaly (Frasnian) coincides with a Big Five mass extinction. The next one (Callovian) is not a major extinction but does mark the fossil-based boundary between the Middle and Upper Jurassic. The next (Cenomanian) is well known as an extinction but has long been associated with evidence of oceanic oxygen depletion (anoxia). The next (Eocene) is a recognized extinction though not a big one; it also contains other evidence of impact. The next (Middle Miocene) matches an extinction peak in Sepkoski's data base but is not widely recognized by paleontologists as a major event. The youngest (Pliocene) is a geographically localized anomaly, and the associated extinction, if there is one, is not large. So, the iridium record is a mixed bag. Whether it supports the impact–extinction link is open to legitimate debate.

Two serious problems dog the iridium data. First, most anomalies were found by searching at known times of extinction. This is understandable because analyses for iridium are expensive and time-consuming. Why not look first where you expect to find it? But the relationship cannot be established firmly without data on the *absence* of iridium at times of no extinctions. More broadly based surveys are being made, but it is too early to say that the iridium-extinction link is firm.

The other problem is that three of the six new iridium anomalies show that the iridium is confined to fossils called stromatolites—microbial mats that trap sedimentary particles. It can be argued, and has been, that the high concentration of iridium is due solely to the organisms' ability to concentrate the trace amounts of the element that occur naturally in seawater. Thus, the iridium may have nothing to do with objects from space. On the other hand, perhaps the organisms carried unusual amounts of iridium because its concentration in seawater was elevated by a meteorite impact.

Are there sizable extinctions without chemical evidence

of meteorite impact? Yes, lots of them. The most obvious is the granddaddy of all extinctions—the Permian—where all efforts to find evidence of impact have failed. The best defense a supporter of the impact-extinction link can make is to point out that many large impacts, especially those of icy comets, need not have associated iridium. Furthermore, Permian rocks with iridium may have existed but were lost to erosion—a reasonable possibility because of the short duration of impact debris deposition.

Turning now to actual craters, one can put together an impressive list of large craters coinciding with extinctions. But one can also compile an unimpressive list. What follows is a pair of advocacy statements, one for each side of the argument.

EXTINCTIONS ARE LINKED TO CRATERS

According to the authoritative summary by Grieve and Robertson (1987), eleven craters with diameters of at least 32 kilometers and ages younger than the start of the Cambrian have been discovered. The 32-kilometer size cutoff is significant because this is the size necessary to cause a 10 percent species extinction, as is shown by the the impact-extinction curve of Figure 10-1. Of the eleven craters, nine are reasonably well dated geologically. Of these, several have geologic ages that virtually match major extinctions.

Of the Big Five mass extinctions, three have large associated craters, as is shown below. The ages given are the most likely (preferred) ages, with uncertainties varying from case to case.

Given the low probability of preservation and discovery of

EXTINCTION CAUSED BY METEORITE IMPACT?

Crater (Diameter)	Age	Age	Mass Extinction
Manson, Iowa (32 km)	65	65	Cretaceous–Tertiary Boundary
Manicouagan, Quebec (100 km)	210	208	Triassic–Jurassic Boundary
Charlevoix, Quebec (46 km)	360		
		367	Frasnian–Famennian Boundry (Devonian)
Siljan, Sweden (52 km)	368		

large craters, it is surprising indeed that as many as three of the Big Five mass extinctions should have matching craters. In fact, the proportion is so high that one must suspect that each mass extinction was caused not by a single impact but by a concentrated shower of objects (probably comets) that increased the likelihood of at least one crater's being preserved. Comet showers have long been accepted as the likely result of disruption of comet orbits by randomly passing stars.

Several lesser extinction events are also matched by crater ages. The giant Popigai Crater in the USSR (100 kilometers) is 39 ma BP, plus or minus 9 ma—an age range that easily includes the terminal Eocene extinction at 35 ma BP. Also, the Clearwater Craters in Quebec (32 and 22 kilometers) formed (simultaneously) at 290 ma BP, plus or minus 20 ma. although the uncertainty of this date is large, the figure of 290 is a perfect match, according to the 1989 Harland time scale, for the extinction that ended the Carboniferous period.

175

EXTINCTION: BAD GENES OR BAD LUCK?

The few large craters not matched by extinctions should be investigated further. The unmatched cases are too few, however, to jeopardize the strong relationship between craters and extinctions.

Small craters, less than 32 kilometers in diameter, while common in the geologic record, are rarely associated with recognized extinctions. A possible exception is Germany's Ries Crater (24 kilometers) and nearby Steinheim Crater (3.4 kilometers). Both are accurately dated at 14.8 ma BP, plus or minus 0.7 ma—close to the small extinction event in the late Middle Miocene (12 ma BP).

Convinced?

EXTINCTIONS ARE NOT LINKED TO CRATERS

Extinction events are fundamentally uncountable, because there is no logical way to split up the continuum of extinction intensity into classes or categories. The Big Five are distinct from the others only by convention. So, whereas it happens that three out of these five appear to have matching craters, the correspondence breaks down when other classifications of extinction are used. If one defines the mass extinctions as the ten largest events, instead of the five largest, the proportion of extinctions matched by craters drops dramatically. Thus, because of the arbitrary nature of extinction classification, unconscious bias cannot be avoided and statistical testing of any conclusions is folly.

The largest extinction of all—the Permian event—carries absolutely no evidence of meteorite impact and therefore must have been caused by other factors.

Impact and extinction dating is notoriously uncertain. Errors in radiometric ages for craters may come from many sources and although plus-minus ranges are usually provided, sources of error and methods of error estimation are rarely given. Even when the stated errors are taken at face value, however, many crater ages have such high uncertainties as to be unusable for comparison with extinction dates. For example, the 360 ma BP age for Charlevoix Crater has a published uncertainty of 25 ma. Thus, Charlevoix could be anywhere between 335 and 385 ma BP, a range that includes several extinctions in addition to the Frasnian event at 367.

If crater dating is uncertain, extinction dating is worse. Consider the date for the end of the Jurassic, an important extinction (Tithonian event) neglected by impact-extinction advocates. In the five major geologic time scales published in the past ten years, the end of the Jurassic has varied from 130 to 145.6 ma BP.

In view of the dating problems, no attempt to correlate ages of craters and extinctions makes sense. Garbage in, garbage out.

Crater size is also a problem because insufficient attention has been given to two large craters that lack associated extinctions. They are the Montagnais Crater, on the continental shelf off Nova Scotia, and the Tookoonooka Crater, in Queensland, Australia. Montagnais is 45 kilometers in diameter and dated at 51 ma BP and Tookoonooka is 55 kilometers and dated at 128 ma BP. Both are considerably above the 32-kilometer cutoff that has been used as the lower limit of significant biological effects. Yet neither has any distinctive extinction associated with it—unless one were to claim a near-match of Tookoonooka with the end-Jurassic extinction (above) by going back to the 1982 Harland time scale.

Even one large crater without an associated extinction is a decisive contradiction. If, as is argued, energy release from a large impact is bound to devastate large numbers of species, then extinction should be associated with all large impacts, not just most or some. This is one instance where conventional statistical reasoning has no place.

From the evidence, we must conclude that the apparent coincidence between craters and extinctions is due to chance, weak sampling, and perhaps bias (undoubtedly unconscious) in the selection of data.

ASSESSMENT

Published research articles in science tend to be advocacy statements. In scientific writing, one seldom admits puzzlement or uncertainty. Rather as in a lawyer's brief, the strongest possible case is made to support each conclusion. I don't know how this got started, but it is part of the culture. Although the practice has some benefits, it has the negative effect of polarizing the scientific community on difficult research issues—issues that do not have a clear answer but which still need discussion and a full exploration of alternative hypotheses.

Depending on my mood, I can support either of the two advocacy statements. So much depends on how one selects and arranges the data. The first statement is attractively upbeat and provides a positive challenge for future research. The second is conservative and negative, and its author is obviously striving to discredit the whole idea. The author of the first seems the nicer fellow, but the second is probably the better scientist, insisting on tight logic and careful verification at every step.

Notice that both authors use a number of tricks and gambits to make their points. The first author, relying on a consensus of the observations, ignores the Montagnais and Tookoonooka craters, whereas the second author emphasizes them as particularly important counterexamples. Perhaps the first author was not aware of these craters—they are not listed in the most recent global compilation of craters (Grieve and Robertson, 1987). The lack of literature citations by both authors is appalling. Finally, the suggestion, made twice by the second author, that the advocates of impact-caused extinction are subject to unconscious bias is a barely concealed accusation of fraud. Innuendo of this sort has no place in the scientific literature.

SOURCES AND FURTHER READING

Azimov, I. 1979. *A choice of catastrophes.* New York: Simon and Schuster. A detailed account of all the things that could go wrong on earth, from collapsing stars to the dangers of over-population.

Calder, N. 1980. *The comet is coming!* New York: Viking Press. A superb account of comets, with special emphasis on the comet Halley; also contains an excellent, early treatment of the Alvarez research.

Grieve, R. A. F., and P. B. Robertson. 1987. *Terrestrial impact structures.* Geological Survey of Canada (Ottawa), Map 1658A. A colorful wall map showing the locations of 116 confirmed meteorite craters on earth.

Sharpton, V. L., and P. D. Ward. 1990. *Global catastrophes in earth history*. Geological Society of America, Special Paper 247. Boulder: GSA. The proceedings of an important conference on extinction held at Snowbird, Utah, in October 1988.

CHAPTER **11**

PERSPECTIVES ON EXTINCTION

HOW TO BECOME EXTINCT

Extinction is a difficult research topic. No critical experiments can be performed, and inferences are all too often influenced by preconceptions based on general theory. There are some things about extinction, however, that we can say with reasonable confidence—things founded on solid observations of fossil and living organisms. I suggest the following:

1. *Species are temporary.* No species of complex life has existed for more than a small fraction of the history of life. A species duration of ten million years is unusually long, yet even this is only about 0.25 percent of

life's tenure on earth. Although some disappearances are due to pseudoextinction—phyletic transformation of one species into another—the prevalence of true extinction is real. The dinosaurs, trilobites, and ammonites are among the many groups in which large numbers of species died without issue.

2. *Species with very small populations are easy to kill.* This follows from the Gambler's Ruin discussion of chapter 3. Also, because species start out small—often as tiny populations on isolated islands—they are near death at the time of their birth. And there is no brooding or parental care for young species. As we noted in Chapter 7, if a species has fewer individuals than its minimum viable population (MVP), extinction in a short time is probable, albeit not assured.

3. *Widespread species are hard to kill.* Species extinction can be accomplished only by the elimination of all breeding populations. Predators must be active over the whole range, not merely most of it. The same is true for extinction caused by competition. If the agent of extinction is a physical disturbance, the killing condition must exist everywhere the species lives.

4. *The extinction of widespread species is favored by a first strike.* The resilience of widespread species can be negated if extreme stress (biological or physical) is applied suddenly over a large area. This is the lesson of the heath hen, described in Chapter 7. The first strike may even be a necessity, but this has not been proven.

5. *The extinction of widespread species is favored by stresses not normally experienced by the species.* Most plants and animals have evolved defenses against the normal vicissitudes of their environment. Although individual organisms live only a short time, successful species live long enough to have experienced—and survived—1,000-year and even 100,000-year events. But a stress that has *never* been experienced by a species can cause extinction. Recall the kill curve, presented in Chapter 4. The larger extinction events occur at average intervals of tens of millions of years, whereas during most shorter time intervals, extinction is negligible. Because Darwinian evolution depends on the *continuous* pressure of natural selection, organisms cannot adapt to conditions they experience only rarely.

6. *The Simultaneous extinction of many species requires stresses that cut across ecological lines.* Many moderate extinction mechanisms are strictly limited to a single ecosystem or habitat. At the extreme is the disease epidemic limited to a single species. Even extinction mechanisms that cause the collapse of a whole ecosystem rarely affect more than one basic habitat. But the larger extinctions seen in the fossil record were clearly more pervasive.

Of the six points just presented, the fifth—that "normal" stresses cannot eliminate widespread species—deserves more discussion. A possible exception is epidemic disease.

Rapid devastations of species over wide areas by disease

are well documented, including attacks on the human species by various plague organisms. However, cases of widespread species actually pushed to complete, global extinction by disease are vanishingly rare—even though it is easy to point to cases, like that of the American chestnut, where species have been pushed *toward* extinction by disease.

The American chestnut was once an important canopy tree in oak-hickory forests of the northeastern United States. In 1906, a fungus (introduced accidentally from China) swept through the region and killed all large chestnut trees. But because the roots of the chestnut are immune to the fungus, the species survived throughout its range. Now, the chestnut is a common constituent of the understory, with trees reaching a height of about twenty feet before being killed by the fungus. Thus, although the case is commonly cited as an example of extinction by disease, the species is still distributed over a large region, maintaining stable populations. It remains to be seen whether the chestnut will continue as an understory species or evolve resistance to the fungus and resume its original role.

The virtual lack of extinctions by disease may stem merely from the fact that human civilization has had only a few thousand years to observe. So, epidemic disease remains a "normal" stress with the unproven potential of killing widespread species.

The notion that slow, gentle pressure produces extinction is part of the Darwinian paradigm. In *The Origin of Species,* Darwin used the metaphor of a log of wood with many wedges driven into its surface. Newly driven wedges were the newly evolved species. With crowding of wedges (species), each new wedge displaced and expelled old ones from the log. The clear implication is that gentle pressure exerted by new—and better-adapted—species leads to the extinction of one or more incumbent species. This idea is appealing and has been learned by generations of biology students. But its verification from actual field data is negligible.

If we accept the foregoing reasoning, we need to search for truly rare and pervasive agents of extinction to explain past extinctions. From among the candidates that have been considered seriously, I favor meteorite impact as the only credible mechanism. Only large impacts have the required energy, are known to have occurred often enough to do the job, yet are rare enough to thwart adaptation by natural selection.

WANTON EXTINCTION

In the 1950s and 1960s, at the height of the cold war, the public was vitally concerned about the effects on health of radioactive fallout. Because of this, good research was done on the physiological consequences of large doses of ionizing radiation. Attention was focused on the immediate effects of radiation, so-called somatic damage, as opposed to longer-term, hereditary effects. The research provided data on somatic damage experienced by all kinds of plants and animals.

For reasons possibly having to do with natural variation in rates of cell division, some organisms are unaffected by radiation doses that kill other organisms outright. For example, insects and grasses have much higher tolerances than mammals.

Now, as a thought experiment, let us imagine that the earth is subjected to a natural dose of high-energy radiation from outer space, perhaps from a nearby supernova. From the research results just mentioned, we can calculate a radiation level that will surely be lethal to all mammals living above ground. But that same dose will leave insects and grasses unscathed. Our calculations need not be especially precise, because of the widely differing susceptibilities of these groups.

Our hypothetical supernova will cause a major event of species extinction. Killing will be highly selective, however; the victims will be organisms for which the radiation dose is lethal, such as terrestrial mammals. Whether all species in the class Mammalia go extinct will depend on the duration of fallout, the ability of burrowing and cave-dwelling mammals to find shelter, and the effect extinctions have on broader ecosystem dynamics—species previously dependent on the victims. In any event, marine mammals (whales, dolphins, and so on) may survive because of the shielding effect of water. Whether or not extinction is complete at the class level, mammals and other susceptibles will be devastated— only a few scattered species with atypical modes of life will remain. But the extinction will pass almost unnoticed by the insects, grasses, and most other nonsusceptibles.

My imaginary scenario, though producing a highly selective extinction, has little or nothing to do with the adaptive quality (fitness) of organisms in normal times. The relative immunity of insects to high-energy radiation is not an adaptation by natural selection, because normal levels of radiation in earth history have been far lower than those of the fallout scenario. Rather, the insects' immunity is a chance by-product of evolution.

This kind of extinction is selective but not constructive; it does not lead to organisms better able to survive in their normal environment—assuming that the postulated radiation levels are too rare in earth history to be grist for natural selection. I call this kind of selective but nonconstructive extinction *wanton* extinction, on the basis of the word's the original meaning: undisciplined or unruly.

The Role of Extinction in Evolution

In the fossil record, many adaptive breakthroughs—bursts of speciation accompanied by the origin of new families and orders—occur after the big mass extinctions. The expansion of mammals immediately following the dinosaur extinction is a classic example. Although this effect is most striking after Big Five mass extinctions, similar patterns are seen at all scales.

In Chapter 1, I suggested that without species extinction, biodiversity would increase until some saturation level was reached, after which speciation would be forced to stop. At saturation, natural selection would continue to operate and improved adaptations would continue to develop. But many of the innovations in evolution, such as new body plans or modes of life, would probably not appear. The result would be a slowing of evolution and an approach to some sort of steady-state condition. According to this view, the principal role of extinction in evolution is to eliminate species and thereby to reduce biodiversity so that space—ecological and geographic—is available for innovation.

A curious exception to extinction-driven evolution is the record of bacteria and other primitive organisms that domi-

EXTINCTION: BAD GENES OR BAD LUCK?

nated the Precambrian (Chapter 2). These organisms do not appear to have changed much during their long tenure on earth. Many of the earliest fossils are indistinguishable in shape and structure from their living counterparts, although there may have been biochemical changes. If these organisms are indeed almost unchanged, perhaps it is because they have never been as prone to extinction as more complex organisms have. Bacteria, in particular, tend to be ubiquitous and to live under extreme conditions. And they are usually very difficult to kill.

NASA and the other agencies around the world that search for extraterrestrial life—especially intelligent life—have recognized the importance of extinction in evolution. Twenty years ago, we thought that stable planetary environments would be best for evolution of advanced life. Now NASA is thinking explicitly in terms of planets with enough environmental disturbance to cause extinction and thereby to promote speciation.

Is selectivity important in the fulfillment of extinction's role in evolution? We have seen, in various contexts in this book, three extinction modes:

1. *Field of Bullets:* random extinction without regard to differences in fitness

2. *Fair game:* selective extinction in a Darwinian sense, leading to the survival of the most fit or best-adapted species

3. *Wanton extinction:* selective extinction, where some kinds of organisms survive preferentially but not be-

cause they are better adapted to their normal environment

All three modes undoubtedly operate at some times and at some scales, but I will argue that the third, wanton extinction, has been the essential ingredient in producing the history of life that we see in the fossil record.

First, however, I must introduce a concept known as *phylogenetic constraint*. It refers to the finding that evolutionary groups tend, in time, toward anatomical stability. Certain growth pathways become genetically fixed. They become so fundamental to the organism that change is rendered improbable or even impossible. The "assembly line" of growth is set up in such a complex way that any significant change calls for a complete redesign, thus constraining evolution to relatively minor variations on existing structures. Because of phylogenetic constraint, it would be difficult for a mammal to add additional legs or make fundamental changes in its digestive system.

If the constraints were completely effective, how could new body plans or new physiologies ever emerge? This question has long puzzled evolutionary biologists, but there are two probable answers. First, innovations in evolution often come from the smaller, simpler, and more generalized members of the ancestral group—from species carrying a minimal load of phylogenetic constraint. Second, bursts of speciation that often follow extinction provide many opportunities, which serves to increase the chance that at least one new body plan or physiology will succeed.

Let me return to the three modes of extinction. In the Field of Bullets mode, large groups of similar and related species—such as the trilobites and ammonites—*will always survive* by virtue of sheer numbers. Suppose, for simplicity, that the world had only two kinds of organisms: red ones and green ones. And suppose that there were ten million species of each kind. Even if 99.9 percent of the twenty million species were killed at random (without regard to color), the chances are overwhelming that many of each color—about 10,000 in this example—would survive. This was the lesson of the trilobite extinction discussed earlier: given many species in the Cambrian and given Field of Bullets extinction, it is mathematically incredible that bad luck alone could have killed off all the trilobites in 325 million years.

The important adaptive radiations of the past were evidently made possible by the disappearance of whole groups of species occupying a range of habitats or modes of life. Thus, extinction must be selective if it is to explain the record.

I am reasonably sure that the required selectivity must be predominantly of the wanton variety. If extinction were always a fair game, with the survivors deserving to survive and the victim deserving to die, we would not have the evolutionary record we see.

I mentioned earlier the history of marine reefs in tropical regions (Figure 2-2). Many different kinds of organisms have, at one time or another, dominated the framework-building role in tropical reefs. The replacement of one of these kinds by another has generally come only after the elimination of all reefs built by the incumbents. If extinction were a fair game, I think the reef community would have settled down early in Phanerozoic time to be dominated by one kind of frame builder—the best one around at the time. This might or might not have been the optimal organism, but its dominance would have discouraged challenges by new and different organisms. In this scenario, we would still have tropical reefs—and they might have worked perfectly well as ecosystems—but much of evolution's variety would never have been realized.

I conclude, therefore, that extinction is necessary for evolution, as we know it, and that selective extinction that is largely blind to the fitness of the organism (wanton extinction) is most likely to have dominated. As Stephen Jay Gould and others have emphasized, we probably would not be here now if extinction were a completely fair game.

Although the foregoing may be true of the role of extinction on earth, a similar biology operating in another planetary environment might evolve differently. Both the Field of Bullets and the fair game modes of extinction are perfectly possible in the appropriate physical environment, even though the results might not be as interesting as those that have occurred here on earth.

BAD GENES OR BAD LUCK?

Extinction is evidently a combination of bad genes and bad luck. Some species die out because they cannot cope in their normal habitat or because superior competitors or predators push them out. But, as is surely clear from this book, I feel that most species die out because they are unlucky. They die because they are subjected to biological or physical stresses not anticipated in their prior evolution and because time is not available for Darwinian natural selection to help them adapt.

Having just made an advocacy statement—bad luck, not bad genes!—I hope the reader appreciates its uncertainties. Favoring bad luck over bad genes is my best guess. It is shared by many of my colleagues even though a majority of paleontologists and biologists still subscribe to the more Darwinian view of extinction, that of a constructive force favoring the most fit species.

Is extinction through bad luck a challenge to Darwin's natural selection? No. Natural selection remains the only viable, naturalistic explanation we have for sophisticated adaptations like eyes and wings. We would not be here without natural selection. Extinction by bad luck merely adds another element to the evolutionary process, operating at the level species, families, and classes, rather than the level of local breeding populations of single species. Thus, Darwinism is alive and well, but, I submit, it cannot have operated by itself to produce the diversity of life today.

A Note on Extinctions Today

Several aspects of this book may seem a bit jarring to readers in this day of heightened concern about endangered species and declining biodiversity. If it is true that well-established species are very hard to kill, why should we be concerned about a little habitat destruction here and a bit of overhunting there? Let me suggest several responses.

The example of the heath hen makes it clear that human activities provide the first strike necessary to reduce species' ranges so that extinction from other causes is likely. Thus, current concerns about endangered species are justified because the human species is producing first strikes regularly—first strikes that nature supplies only at intervals of millions of years.

Much of the current concern about extinction and declining biodiversity is predicated on the conviction that *all* species are important and must be protected. This conviction has been justified on a variety of moral, aesthetic, and pragmatic grounds; all are valid from a humanistic standpoint. If one is committed to protection of all species, then the paving over of a few acres of land for a parking lot is important. Because most species have small ranges, large numbers of species will inevitably be eliminated by localized habitat destruction, just as natural calamities eliminate localized species. But my purpose in this book has been to concentrate on the major players over the billions of years of the history of life, and this has led inevitably to greater emphasis on the widespread species and their surprising resistance to extinction.

EPILOGUE:

DID WE CHOOSE
A SAFE PLANET?

The fossil record documents extinctions of many species that were doing fine—until their demise. Are we any different? Is *Homo sapiens* vulnerable to a first strike of natural origin that could do us in despite our proven ability to cope with the normal vicissitudes of nature? Or could a first strike of lesser magnitude—short of total extinction—devastate human civilization as we know it?

Whatever our view of meteorite impact as a cause of extinction in the past, the risk of comet and asteroid impact is part of our present-day environment. But is the risk great enough, compared with other daily hazards, to justify concern? If so, is there anything we can do about it?

Considerable effort has gone into answering these questions. In 1981, the Jet Propulsion Laboratory sponsored a conference at Snowmass, Colorado, chaired by Gene Shoe-maker and attended by some of the best solar system astron-omers, astrogeologists, engineers, and aeronautics experts. A report of the Snowmass meeting was drafted but never re-leased, apparently because the principals were too busy with other projects, including searches for more asteroids in earth-crossing orbits. Fortunately, a firsthand summary may be found in the final chapter of *Cosmic Catastrophes,* by Clark Chapman and David Morrison.

What are the chances that earth will be hit by a damaging asteroid or comet in our lifetime or the lifetimes of our children or grandchildren? On the one hand, no human fatalities from meteorite strikes have been reported, thus suggesting that the risk is negligible. On the other hand, the Tunguska event of 1908 (see chapter 9) could have wiped out a major city. Its energy was the equivalent of twelve megatons of TNT—about a thousand times the current fig-ure for the atom bomb dropped on Hiroshima (thirteen kilo-tons). Isaac Asimov calculated that if the Tunguska comet had hit six hours later—a quarter turn of the earth—it might have eliminated the city of St. Petersburg (Leningrad).

Tunguska-size (or larger) events are estimated to occur, somewhere on earth, every 300 years (average waiting time). If this estimate is correct, more than a dozen Tungus-kas have occurred during recorded human history. That only one of them is known to us is not surprising, since so much of the earth's surface (including oceans) is un-populated. Thus, the actual waiting time between city-de-stroying events is far greater than 300 years.

But there is another approach to risk assessment. The Snowmass participants concluded that a "civilization-de-stroying" impact occurs, on average, every 300,000 years.

Such an impact would release energy equivalent to about 100,000 megatons—more than 8,000 times that of Tunguska and 8 million times that of Hiroshima. "Civilization-destroying" means devastation so great that a long period comparable to the Dark Ages (or possibly the Stone Age) would be likely.

If the average spacing of the event is 300,000 years, then the odds are 1:300,000 of civilization's destruction in a given year. For an person living for seventy-five years, the lifetime risk is 1:4,000. This gets into the range of other natural and man-made hazards. Chapman and Morrison, in *Cosmic Catastrophes,* provide some intriguing comparisons. They note that one's chances of experiencing a civilization-destroying impact sometime during a normal lifetime are substantially greater than the chances of dying in an airplane crash, about the same as death through accidental electrocution, and about one-third the risk of being killed in an accidental shooting.

In pondering risk assessment, the Snowmass participants were careful to point out how little really reliable information is available on impact probabilities. Although the best estimate of the chance of a civilization-destroying event is 1:300,000 per year, the Snowmass group concluded that the actual odds could be anywhere between 1:10,000 and 1:1,-000,000. The uncertainty is due partly to our incomplete knowledge of comet and asteroid populations (and their orbits) and partly to ignorance of environmental effects.

The uncertainty is far greater than the factor-of-two estimate given by Gene Shoemaker for the timing of large craters (Chapter 9). Ironically, estimates are more dependable for large craters than for smaller impacts because the small ones do not leave a lasting geological record.

197

Clearly, the risk to human populations may be seen as trivial or significant, depending on what one considers a serious disaster and how one works through the arithmetic. My own feeling is that it is too close to call. I don't know on which side of the boundary between acceptable and unacceptable risk we lie. Chapman and Morrison are also cautious. They note that the Snowmass estimate makes the risk (per person) of a devastating impact *greater* than the risk of death from "TCE [trichloroethylene, a carcinogen], asbestos insulation, saccharin, fire-crackers, or nuclear power"— yet far *less* than the risks of death from cigarettes or automobile accidents.

In all this, much depends on whether impact is avoidable. And this reduces to several questions: Can there be warning of an impending impact? If so, could anything be done to prevent impact? Would there be time enough?

It is estimated that only about 5 percent of large (> 1 km) asteroids with earth-crossing orbits have been discovered. For these, orbits are known well enough that we should have warning times of several years or even decades. But for the 95 percent of earth-crossers that have not been discovered, warning times would be very short, and would depend upon actual sightings of an approaching body.

In March of 1989, earth experienced one of many near-misses. An asteroid at least one-third of a kilometer in diameter passed at a distance about twice that from here to the moon. This asteroid, named 1989FC, was not noticed until several days *after* it passed earth. The normal method of detecting asteroids is to compare photographs of the sky-scape taken about forty-five minutes apart. After adjustments for the normal displacement of stars due to the earth's rotation, an asteroid (or comet) is noticeable as a point of light that has shifted between the times when the photographs were taken. Ironically, if the object is coming directly toward earth, no displacement will be seen and the object will probably not be noticed.

Thus, warning time may not be enough for useful response or reaction. Some people are optimistic, however. In May of 1990, the American Institute of Aeronautics and Astronautics (AIAA) issued a position paper arguing strongly that serious study (and funding) be given to the impact problem—to increase knowledge of solar system objects with earth-crossing orbits and to explore "methods and technologies for deflecting or destroying" the objects when impact is predicted. The AIAA statement was endorsed by Vice-President Quayle in his capacity as chairman of the National space Council. Again, whether one considers the case to be strong or weak depends on one's assessment of the accuracy of the current estimates of impact probability.

It is a catch-22 situation: we will not know how safe (or unsafe) our planet is until more research is done on collision rates—but funding for that research is hard to justify without stronger evidence that the societal risk of impact is serious. We don't know whether we chose a safe planet.

SOURCES AND FURTHER READING

Chapman, C. R., and D. Morrison. 1989. *Cosmic catastrophes.* New York: Plenum Press. A well-written and authoritative account of the earth's place in the universe, with a good discussion of the present-day risks of a serious meteorite impact.

Morrison, D., and C. R. Chapman. 1990. Target earth: It *will* happen. *Sky and Telescope*, March, 261–65. Further discussion of the likelihood of a serious impact.

Trefil, J. S. 1989. Craters, the terrestrial calling cards. *Smithsonian*, September, 80–93. An excellent treatment of comets and asteroids and their impact craters.

NATURE BATS LAST. *Sheila O'Hara,* © *1989*

Sheila O'Hara, born in Japan and trained in fine arts in California, is recognized internationally for her innovative tapestries. With her poignant style, loaded with deep irony and humor, she has attracted major commissions from corporations and individuals in the United States and abroad. She resides at 7101 Thorndale Drive, Oakland, CA 94611.

INDEX

INDEX

bacteria, 22–23, 24, 187–88
Barringer Crater (Meteor Crater),
    157–58, 159
bats, 52, 53
bell-shaped curves (Gaussian
    distributions), 57, 58
Bermuda, 116
Biarritz, 78
Big Five mass extinctions, 65–70, 67,
    81–83, 187
    kill curve and, 85
    list of, 65
    meteorite impact and, 158, 162–63,
        164, 172, 174–75, 176
Billingham, J., 155
biodiversity, xii, 11, 24, 54
    Cambrian explosion and, 26–27
    Devonian, 32
    rain forests and, 135
    species extinction and, 18–20, 18, 187
    species rarity and, 59–60
    tropical reefs and, 36
biological causes of extinction, 118–38
    competition and, 127–29
    first strike importance and, 122–23, 182
    fragility of species and ecosystems and,
        119–20
    great American interchange and,
        133–34
    heath hen and, 121–23, 125
    history of tropical rain forests and,
        134–37, 136
    small populations and, 124–27
    species-area effects and, 129–34, 130
    see also causes of extinction; physical
        causes of extinction
biology, biologists, 40–41
    conservation, 124, 129–34, 130, 144–49
    interest in extinction and, 9–12, 11
    tropical reefs and, 36–38, 36–37
biostratigraphic zone, 170
birds, 33, 34, 90
    competition of, 128–29
    dinosaurs' relationship to, 7, 8
    taxonomy and, 16

birthrate, 74–75, 125
blitzkrieg theory, 89–93
    critics of, 92–93
body size, selectivity of extinction and,
    91–95
bolide, 162
breeding experiments, 15, 16
broken-stick model, 58, 59
Burgess Shale, 25–26, 31

calcium carbonate, 26
Calder, N., 179
Callovian iridium anomaly, 172
Cambrian explosion, 26–27
Cambrian period, 25–29, 30, 31, 67,
    143
    trilobites in, 25, 102
    tropical reefs in, 37
camels, 90
cancer, skewed variations and, 57
carbon-14 dating, 89, 90, 92
Carboniferous period, 29, 31, 37, 67, 143,
    175
Caribbean, crater search in, 163
casino analogy, 45–49, 47, 51–52
causes of extinction, 107–55, 181–85
    anthropomorphism and, 112–14
    Just So Stories and, 110–12
    kill curve and, 114–17, 115
    rarity of extinction and, 107–10
    search for, 107–17
    species-specific, 116–17
    see also biological causes of extinction;
        physical causes of extinction
Cenomanian iridium anomaly, 172–73
Chapman, Clark, 196–98, 199, 200
Charlevoix Crater, 177
chemical poisoning, of oceans, 113
chemistry:
    of atmosphere, 113
    origin of life and, 23
    Precambrian, 24–25
chestnut, American, 184
China, 78, 89
civilization-destroying impact, 196–97

INDEX

206